Water Transport in Plants under Climatic Stress

Water Transport in Plants under Climatic Stress

edited by
M. Borghetti
Consiglio Nazionale delle Ricerche, Firenze, Italy
(now at the University of Basilicata, Italy)

J. Grace
University of Edinburgh, UK

and
A. Raschi
Consiglio Nazionale delle Ricerche, Firenze, Italy

**Proceeding of an International Workshop, held in
Vallombrosa, Firenze, Italy**

CAMBRIDGE UNIVERSITY PRESS
Cambridge, New York, Melbourne, Madrid, Cape Town, Singapore, São Paulo, Delhi

Cambridge University Press
The Edinburgh Building, Cambridge CB2 8RU, UK

Published in the United States of America by Cambridge University Press, New York

www.cambridge.org
Information on this title: www.cambridge.org/9780521112000

First published 1993
This digitally printed version 2009

A catalogue record for this publication is available from the British Library

ISBN 978-0-521-44219-0 hardback
ISBN 978-0-521-11200-0 paperback

Contents

viii   Contents

List of contributors

T. Ameglio
INRA, Domaine de Crouelle, 63039 Clermont-Ferrand cedex, France.

L. Angelocci
ESA Luiz de Queiroz, USP, Caixa Postal 9, 13400 Piracicaba, SP, Brasil.

A.V. Anisimov
Kazan Institute of Biology, 42111, P.O.B. 30, Kazan, Russia.

M.N. Andersen
Department of Soil Tillage, Soil Physics and Irrigation, Danish Research Service for Plant and Soil Science, Flensborgvej 22, DK-6360, Denmark.

G. Bacic
Dept. of Phys. Chem., Faculty of Science, University of Beograd, Beograd, Yugoslavia.

A. Berger
CNRS - CEFE BP 5051, 34033 Montpellier cedex, France.

M. Borghetti
Istituto Miglioramento Genetico delle Piante Forestali, Consiglio Nazionale delle Ricerche, via S. Bonaventura 13, 50145 Firenze, Italy. Present address: Dipartimento di Produzione Vegetale, University of Basilicata, via. N. Sauro 85, 85100 Potenza, Italy.

G.J. Campbell
Botany Department, University College, Dublin Belfield, Dublin 4, Ireland.

Čermák
Institute of Forest Ecology, University of Agriculturae, Brno, Czechoslovakia.

Y. Cohen
Department of Agricultural Meteorology, A.R.O., The Volcani Center, Bet Dagan, Israel.

P. Cruiziat
INRA, Domaine de Crouelle, 63039 Clermont Ferrand cedex, France.

P. De Angelis
Dipartimento Scienze dell'Ambiente Forestale e delle sue Risorse, University of Tuscia, via S. Camillo De Lellis, 01100 Viterbo, Italy.

A. Del Lungo
Dipartimento Scienze dell'Ambiente Forestale e delle sue Risorse, University of Tuscia, via S. Camillo De Lellis, 01100 Viterbo, Italy.

M.A. Dixon
Department of Horticultural Science, University of Guelph, Guelph, Ontario, Canada, N1G 2WI.

D.J. Durzan
Department of Environmental Horticulture, University of California, Davis, CA 95616-8587, USA.

I. Ferreira
DER, Instituto Superior de Agronomia, Universidada Técnica de Lisboa, Papada da Ajuda, 1399 Lisboa Codex, Portugal.

S. Frisullo
Dipartimento di Patologia vegetale, University of Bari, sede di Foggia, Foggia, Italy.

M. Fuchs
Department of Agricultural Meteorology, A.R.O., The Volcani Center, Bet Dagan, Israel.

J. Grace
Institute of Ecology and Resource Management, The University of Edinburgh, Darwin Building, Mayfield Road, Edinburgh EH9 3JU, UK.

A. Granier
INRA, Station de Sylviculture et Production, Laboratoire d'Ecophysiologie et de Bioclimatologie, BP 35 Champenoux, 54280 Seichamps, France.

K. Gross
Institute of Silviculture, University of Freiburg, Bertoldstr. 17, D-7800 Freiburg, Germany.

J. Heimann
Institut für Forstbotanik, University of Göttingen, Büsgenweg 2, D-3400 Göttingen, Germany.

H. Heydt
University of Bayreuth, Universitätsstraße 30, D-8580 Bayreuth, Germany.

P.G. Jarvis
Institute of Ecology and Resource Management, The University of Edinburgh, Darwin Building, Mayfield Road, Edinburgh EH9 3JU, UK.

C.R. Jensen
Dep. Agr. Sci., Section of Soil and Water and Plant Nutrition, The Royal Veterinary and Plant Nutrition, DK-1871 Frederiksberg C., Denmark.

R.W. Johnson
Department of Horticultural Science, University of Guelph, Guelph, Ontario, Canada, N1G 2WI.

H.G. Jones
Horticulture Research International, Wellesbourne, Warwick CV35 9EF, UK.

W. Koch
Institute of Forest Botany, University of München, Amalienstr. 52, D-8000 München 40, Germany.

J. Kucera
Institute of Forest Ecology, University of Agriculturae, Brno,
Czechoslovakia.

M.A. Lo Gullo
Istituto di Botanica, Università di Messina, v. P. Castelli 2, 98100
Messina, Italy.

R. Lösch
Abt. Geobotanik, Universitätsstraße 1/26.13, D-W-4000 Düsseldorf,
Germany.

D. Loustau
INRA Station de Recherche Forestiers, Laboratoire d'Ecophysiologie et
Nutrition, BP. 45 Gazinet, 33610 Cestas, France.

J.A. Milburn
Department of Botany, University of New England, Armidale, NSW
2351, Australia.

S. Moreshet
Department of Agricultural Meteorology, A.R.O., The Volcani Center,
Bet Dagan, Israel.

N. Moretti
Dipartimento di Produzione Vegetale, University of Basilicata, via N.
Sauro 85, 85100 Potenza, Italy.

Z. Nasr
INRA Laboratoire de Bioclimatologie, Centre de Bordeaux, Domaine de
la Grande Ferrade, BP 81 F3388 Villenave d'Ornon, France.

J.J. Oertli
Institute of Plant Science, ETH, 8092 Zürich, Switzerland.

B.A. Osborne
Botany Department, University College,
Dublin Belfield, Dublin 4, Ireland.

A. Pardossi
Dipartimento di Biologia delle Piante Agrarie, Sezione di Orticoltura e Floricoltura, Università di Pisa, Viale delle Piagge 23, 56100 Pisa, Italy.

J.A. Petty
Department of Forestry, University of Aberdeen, St. Machar Drive, Aberdeen AB9 2UD, UK.

M. Pisante
Dipartimento di Produzione Vegetale, University of Basilicata, via N. Sauro 85, 85100 Potenza, Italy.

A. Raschi
Istituto di Analisi Ambientale e Telerilevamento applicati all'Agricoltura, Consiglio Nazionale delle Ricerche, p.le delle Cascine 18, 50144 Firenze, Italy.

S. Ratkovic
Department of Technology and Chem. Res., Maize Research Institute, 11080 Zemum-Beograd, Yugoslavia.

D.J. Robson
The BioComposites Centre, University of Wales, Bangor, Gwynedd, LL57 2UW, UK.

M. Sabatti
Dipartimento Scienze dell'Ambiente Forestale e delle sue Risorse, University of Tuscia, via S. Camillo De Lellis, 01100 Viterbo, Italy.

S. Salleo
Istituto di Botanica, Università di Messina, via P. Castelli 2, 98100 Messina, Italy.

G.E. Scarascia Mugnozza
Dipartimento Scienze dell'Ambiente Forestale e delle sue Risorse, University of Tuscia, via S. Camillo De Lellis, 01100 Viterbo, Italy.

G. Selles
Comision Nacional de Riego. Teatinos 50/5to Piso, Santiago, Chile.

J. Sperry
Department of Biology, University of Utah, Salt Lake City, UT 84112, USA.

E. Steudle
Universität Bayreuth, Universitätsstraße 30, D-8580 Bayreuth, Germany.

W. Stickan
Systematisch-Geobotanisches Institut, Universität Göttingen, Untere Karspüle 2, D-3400 Göttingen, Germany.

R.A. Sutherland
Horticultural Research International, Wellesbourne, Warwick CV35 9EF, UK.

R. Tognetti
Istituto Miglioramento Genetico delle Piante Forestali, Consiglio Nazionale delle Ricerche, via S. Bonaventura 13, 50145 Firenze, Italy.

F. Tognoni
Dipartimento di Biologia delle Piante Agrarie, Sezione di Orticoltura e Floricoltura, Università di Pisa, Viale delle Piagge 23, 56100 Pisa, Italy.

C. Valancogne
Laboratoire de Bioclimatologie, INRA, Centre de Bordeaux, Domaine de la Grande Ferrade, BP 81 F3388 Villenave d'Ornon, France.

R. Valentini
Dipartimento Scienze dell'Ambiente Forestale e delle sue Risorse, University of Tuscia, via S. Camillo De Lellis, 01100 Viterbo, Italy.

P. Vernieri
Dipartimento di Biologia delle Piante Agrarie, Sezione di Orticoltura e Floricoltura, Università di Pisa, Viale delle Piagge 23, 56100 Pisa, Italy.

Editors' preface

An important International Workshop was held at the Vallombrosa Abbey, in the Forest of Vallombrosa, near Firenze, Italy, 29-31 May 1990. Eighty scientists participated in a discussion of water transport in plants. There have been many international workshops and conferences on plant-water relations but this was the first to focus on the failure of the hydraulic pathway within the xylem. It was possible to assemble practically all those scientists, worldwide, who have worked on the cavitation of water in the transport system. This phenomenon of cavitation, which was discovered only in the 1960s, is now being recognized as being widespread. It occurs in all the species of vascular plant so far examined, and can usually be detected on any summer's day. Its ecological significance is a matter for further research, but many consider that embolism in the xylem predisposes plants to further water stress, so that cavitation and refilling may hold the key to vegetational response to climatic warming and drying.

At the meeting it was resolved to prepare a manuscript for publication and this process (with peer review and revision) took place during 1991.

Papers presented fall naturally into several subject groupings:
(i) analysis of the mechanism and pathway of water flow in the plant,
(ii) the natural repair of the hydraulic continuum, whereby emboli are redissolved in water,
(iii) survey of methodologies including acoustic detection of cavitation, thermoelectric techniques and nuclear magnetic resonance, and
(iv) case studies, examples of current work, mainly in the hot dry climates of the southern Mediterranean.

The workshop was sponsored by local, national and European organisations, without which the meeting would not have taken place. These were the Commission of the European Community, the Consiglio Nazionale delle Ricerche, the Society of Experimental Biology, the British Ecological Society and the Centro Studi per l'Informatica in Agricoltura (Firenze). In addition we received valuable support from the Vallombrosa Abbey, Comune di Reggello and Fondazione Scienza per l'Ambiente.

We are particularly gratefull to S. Bianchi, P. Cioni, J. Gallori and F. Giannini who worked to ensure the smooth running of the workshop and, finally, to Dr. R. Tognetti who assisted us tirelessly in the editorial work.

Global change and plant water relations

P. G. JARVIS

Institute of Ecology and Resource Management,
University of Edinburgh, Darwin Building, Mayfield Road,
Edinburgh EH9 3JU, UK.

SUMMARY

Physiological responses to elevated CO_2 are discussed at leaf, plant and stand scale in the context of global change and their consequences for water relations at these scales evaluated. A distinction is drawn between short term responses and the responses shown by plants that are fully acclimated by extended periods of growth in a high CO_2 environment.

Assimilation of CO_2 and stomatal action are the best known processes dependent on CO_2 at leaf scale and acclimation reduces their impact on growth, transpiration and water use efficiency.

At plant scale leaf and root growth are generally stimulated in high CO_2 but the processes involved are poorly understood. The consequence may be increase in rate of both transpiration and water uptake. Control system based models explicitly including feedbacks offer a means of integrating multiple interacting limiting variables and of analysing sensitivity of processes to increase in CO_2 concentration at both leaf and plant scale. To make use of such models to give helpful predictions, better definition of "pressure points" - the processes on which CO_2 is known to act - is needed. Response of processes within the plant to elevated atmospheric CO_2 is strongly influenced by coupling between leaves and atmosphere.

At the larger stand scale, this should be taken into account explicitly because the degree of coupling depends on the structure of the vegetation. Increase in leaf area will have a larger effect on transpiration from well coupled vegetation (*e.g.* tall crops, shrubs and trees) but may have little effect on transpiration from poorly coupled vegetation (*e.g.* short crops, grass lands, dwarf shrubs).

Empirical data provide little guide as to the likely effects of increase in atmospheric CO_2 concentration on plant water relations but overall it seems that the effects may be either positive or negative, and will probably be small.

INTRODUCTION

The global atmospheric CO_2 was in a steady state at a concentration of about 280 μmol/mol 150 years ago and has been rising steadily since then as a result of man's activities. Over the past 20 years the annual rate of increase has varied, depending on global economics, between 1 and 2 μmol/mol and is currently about 1.8 μmol/mol per annum (Houghton *et al.*, 1990).

Up until the Second World War, the main new source of atmospheric CO_2 was the oxidation of carbon in soils and vegetation as more land was brought into cultivation and burning of vegetation became more frequent. In the middle of this century, changes in land use were overtaken by the exponentially increasing use of fossil fuels, particularly oil and gas, as the main source of anthropogenic CO_2. Today the use of fossil fuels puts about 5.7 Gt/a into the atmosphere and land use changes, essentially tropical deforestation, add another 2 to 3 Gt/a (Hammond, 1990; Houghton, 1991). Of this about 3.3 Gt/a remains in the atmosphere, accounting for the rise in atmospheric concentration (Houghton *et al.*, 1990), the remainder being absorbed by oceanic and terrestrial biota, whence it passes into storage (Tans *et al.*, 1990).

This rise in atmospheric CO_2 concentration is *likely* to affect global climate and to cause regional changes in air temperature and humidity, length of growing season, precipitation and evaporation, all of which will affect plant water relations.

However, so far we can not say with certainty that such changes are occurring globally and predictions of such changes on a regional scale are presently very unsafe (Houghton *et al.*, 1990). It may take another 10 years or more to establish that the existing trends in climate data do indeed indicate that global warming is occurring, and a similar period of development of global climate models to provide reliable regional predictions of climate change.

We do know, however, that the atmospheric CO_2 concentration has been rising and we can be fairly certain that it will continue to rise with present energy policies for many years to come. Current forecasts based on the "Business as usual" scenario suggest a concentration of at least 500 μmol/mol by the year 2050 (Parry *et al.*, 1991).

Carbon dioxide has marked physiological effects on plants, some of which we know very well as the result of a long history of scientific endeavour in photosynthesis, and others about which we are much less certain and which thus provide a focus for renewed study.

In the following paragraphs, I shall, therefore, briefly summarise effects of CO_2 on leaves, plants and vegetation as it may affect water relations, and leave more speculative effects of changes in regional climate aside for the time being.

PHYSIOLOGICAL EFFECTS OF CO_2

It has been known for many years that raising the ambient CO_2 concentration enhances plant growth and wide use of this has been made in the horticultural industry to bring certain crops, such as lettuce or carnations, to an earlier harvest. In the earlier work pertaining to this goal, CO_2 was added to give quite large and variable concentrations so that the results are not particularly helpful with respect to, say, a likely doubling in atmospheric CO_2 concentration by the middle of the next century, but they have been helpful in defining experimental programmes that have developed over the last 10 years. Still, however, the majority of experiments reported are inadequate with respect to a number of features including, particularly, poor experimental design and lack of adequate replication, as well as little or no control of nutrition, constrained rooting in pots, etc. (see Eamus & Jarvis (1989) for a critical discussion of methodological constraints).

For herbaceous and woody C3 and C4 plants, growth is enhanced by a doubling of the ambient CO_2 concentration. Growth increases reported in the literature range from 0 to 200% with a median of about 40% (Acock & Allen, 1985; Eamus & Jarvis, 1989). Whilst some of this variation may be due to differences amongst species, much of it may well result from differences and inadequacies amongst the experimental protocols.

There are three main reasons for growth enhancement of plants by elevated ambient CO_2 concentration: effects on photosynthesis, on stomatal action and on leaf growth. In photosynthesis, CO_2 is not only the substrate for carboxylation of RuBP by Rubisco, but it also competes for the active sites on Rubisco with oxygen (*i.e.* it reduces or inhibits photorespiration) and it is an activator for Rubisco (Sage *et al.,* 1988). Whilst knowledge of the rôle of CO_2 as an activator is relatively new, its primary rôle in photosynthesis has been known and understood for many years. The CO_2 assimilation rate (A) increases with the mean intercellular space CO_2 partial pressure (or mole fraction) (C_i) which is a function of the ambient CO_2 mole fraction (C_a), and is dependent on the rate of assimilation of CO_2 by the chloroplast and the conductance of the stomata, and the coupling of the leaf to the ambient air (Fig. 2). Similarly, it has been known for a long time that a high ambient CO_2 concentration reduces stomatal aperture and conductance (g_s) (Meidner & Mansfield, 1965). Although the mechanism by which this is effected is still unclear, the stomata also respond primarily to C_i (Mott, 1988). Thus changes in C_a lead to changes in C_i which in turn affects both g_s and A.

The feedbacks amongst these processes are complex (Fig. 1) but the overall operational responses are such that C_i/C_a is conservative, remaining approximately constant unless deliberately manipulated by experimentation (Jarvis & Morison, 1981; Morison, 1987). Thus C_i can be regarded as the effective sensor for both A and g_s in respect of a change in ambient CO_2 mole fraction (Mott, 1990).

Early studies of leaf energy balance (Raschke, 1958) also indicated that a rise in leaf temperature resulting from stomatal closure would partially offset the effect of stomatal closure on transpiration. Again, the feedbacks amongst the processes are complex (Fig. 1), a rise in leaf temperature increasing radiative transfer, respiration rate and saturation deficit at the leaf surface, with primary consequences for net radiation, C_i and g_s.

In general, elucidation of the fundamental effects of CO_2 on particular processes has been achieved by taking plants growing in the CO_2 environment current at the time and subjecting them to *short term* exposure to different ambient CO_2 concentrations and analysing their responses. It has become evident in the last 10 years, however, that growing plants for extended periods of weeks, months or years in elevated atmospheric CO_2 concentration leads to adjustment of the rates of processes and this *acclimation* must be taken into account when analysing the consequences of a rise in atmospheric CO_2 concentration on growth and water relations. Full acclimation may require growing plants in a high CO_2 environment continuously for more than one generation.

ACCLIMATED RESPONSES TO ELEVATED CO_2 CONCENTRATION

The degree of acclimation to elevated ambient CO_2 seems to depend on two things: the availability of sinks in the plant for the carbon (C) assimilated, and the availability of nitrogen (N) at the root surface for uptake to balance the enhanced availability of carbon. Making C more freely available by increasing the ambient CO_2 concentration seems to have much the same developmental effect as limiting the supply of N to plants in the current CO_2 concentration: unless the greater availability of C is balanced by greater availability of N, developmental changes characteristic of N deficiency occur. Most apparent, and evident in very many experiments, allocation of C to the roots is enhanced, and a larger root system results, both fine and coarse root growth being increased. If, however, N is made freely available at the root surface, by frequent additions, allocation of C may not be affected.

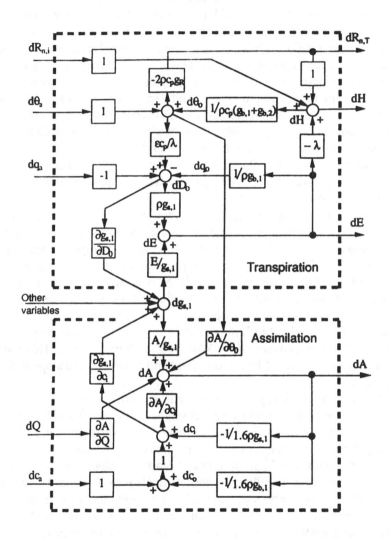

FIG. 1. Control diagram showing the interrelated responses of transpiration and assimilation of a hypostomatous leaf to an environment where any of a number of inputs may vary. The 'other variables', which may affect stomatal conductance directly, include changes in soil water content and quantum flux density. The effect of changing boundary-layer conductance is not included in this diagram. It should be noted that the conventional positive directions for A and E are opposite, necessitating a negative gain, $-1/1.6\rho g_b$, in the C_0 feedback pathway (from McNaughton & Jarvis, 1991).

Thus this appears to be a response to maintain a constant C/N ratio in the plant. In circumstances in which phosphorus is a scarce resource, a similar interaction between C and P is evident (*e.g.* Conroy *et al.,* 1990).

Acclimation of the photosynthetic system in plants grown at elevated CO_2 is evident in CO_2 assimilation at leaf scale as a reduction in the slope of the relationship between A and C_i. This reduction in the A/C_i function can be correlated with a reduction in Rubisco activity, sometimes associated with a reduction in amount and sometimes with a reduction in activation of Rubisco, and with reduction in limitation by regeneration of orthophosphate (Sage *et al.,* 1989). As a result, the overall stimulation of CO_2 assimilation is often much less than would be anticipated from short term responses alone (Fig. 2). This acclimation in Rubisco activity takes place over a period of 5 to 10 days and is apparently reversible on a similar time scale.

The reduction in slope of the A/C_i function may be interpreted in more than one way. One hypothesis hinges on the balance between availability of CO_2 and of nitrogen. Typically between $1/3$ and $1/2$ of the nitrogen in a leaf is committed to Rubisco. If carbon becomes more readily available as a result of CO_2 fertilisation and if N is a scarce resource, as it frequently is, optimisation of the functioning of the photosynthetic system would be expected to result in a redistribution of N from Rubisco to the enzymes involved in light harvesting and downstream processing of the products of carboxylation (Woodrow & Berry, 1988). This then would be seen as a loss of Rubisco activity and a reduction in slope of the A/C_i function.

Alternatively, if the capacity for growth is limited and adequate sinks for C are not available, carboxylase activity is reduced and a reduction in slope occurs. If, however, there are adequate active sinks present, such as growing fruits, there may not be any reduction in the slope of the A/C_i function. It seems that in some cases at least, the slope reduction may result from feedback inhibition because of inadequate C sinks. However, at the present time, there is considerable uncertainty as to the generality of a change in the A/C_i function in response to elevated CO_2 concentration, particularly in different root growth environments.

The stomatal conductance response function (g_s/C_i) is similar in form in plants grown at elevated CO_2 and at the current CO_2 concentration but acclimation also occurs and may result in displacement of the response curve upwards or downwards leading to either higher or lower stomatal conductance at the same C_i. In the former case the value of g_s may be very similar at the two CO_2 growth concentrations, but the latter case of a further reduction in g_s at high CO_2 concentrations seems to be a more common response.

Similarly, stomata of elevated CO_2 plants show typical response functions with respect to photon flux density and saturation deficit but quantitative differences in response, relative to plants grown in the current CO_2 concentration, have also not been well defined. A major limitation to progress in this area, is that the fundamental action of CO_2 in stomatal functioning is not understood at present.

The combined effects of acclimation of A and of g_s with respect to CO_2 enrichment are shown in Fig. 2. It is evident that the overall enhancement of assimilation rate in a high CO_2 environment may vary widely depending on the reduction in the A/C_i function and the direction and magnitude of stomatal acclimation.

Effects of CO_2 on leaf growth are also not well understood at present. It is an observation common to many experiments, and indeed to CO_2 fertilisation in horticulture, that plants grown in elevated CO_2 have large leaves and reach both maximum leaf size and maximum plant leaf area earlier than plants grown in the current CO_2 concentration (*e.g.* Morison & Gifford, 1984; Radoglou & Jarvis, 1990).

Clearly, this might simply be the result of enhanced availability of carbohydrate substrates for leaf growth but other possibilities such as enhanced turgor, hormonal effects and direct effects on cell wall properties and cell enlargement merit investigation. Cell size in leaves is often increased in elevated CO_2 and this might result from higher solute potential and turgor pressure, but increased wall loosening and other possible changes need to be considered. In perennial shrubs and trees the phenology of bud burst and bud formation may also be influenced, as well as the longevity of individual leaves. Thus, it is a distinct possibility that enhanced growth in elevated CO_2 may result as much from effects on dynamics of the leaf population as on photosynthesis and stomatal action. Thus a crucial question relates to what is cause and what is effect amongst leaf growth, assimilation rate and plant growth. Furthermore, increase in leaf area may well offset any savings of water resulting from stomatal closure.

As with carbon assimilation, we may also analyse the impact of elevated CO_2 on both the demand and supply functions for water. So far, we have considered only the demand side of the equation - the variable plant properties of stomatal conductance and leaf area that determine transpiration. Since, however, plant development depends strongly on the balance between assimilation of C and N and their subsequent partitioning into new growth, it is to be anticipated that elevated CO_2 will also result in changes to properties of the water uptake and supply system.

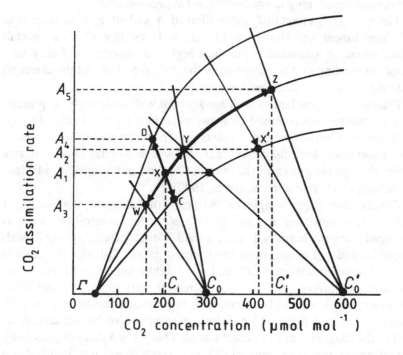

FIG. 2. A diagram to show the possible consequences of an increase in ambient CO_2 concentration for leaf CO_2 assimilation rate, taking into account likely effects on stomatal conductance. The line ΓZ is the A/C_i demand function. The line C_0D depicts the supply function that intersects at point X to give the assimilation rate A_1. The slope of this line is the stomatal conductance. If the stomata open or close somewhat the intersection moves to Y or W giving the new assimilation rates of A_2 or A_3, respectively. If Rubisco activity increases or decreases then assimilation moves onto the lines ΓD or ΓC, respectively, with higher (A_4) or lower assimilation rates. A rise in the leaf surface CO_2 concentration (from C_0 to C'_0) to 600 μmol mol^{-1} with no change in Rubisco activity or stomatal conductance would increase the assimilation rate to A_5. Acclimation in both could bring the assimilation rate back to A_1, or anywhere within the area defined by C'_0Y, C'_0Z and YZ. The effect of changing boundary layer conductance is not included in this diagram (Jarvis, 1989).

Increases in root mass and length have been found in a number of experiments but so far little definitive evidence of changes in the vascular systems of herbs or of trees has emerged.

EFFECTS OF HIGH CO_2 ON WATER RELATIONS AT LEAF AND PLANT SCALE

In comparison with plants growing in today's CO_2 environment, at *leaf scale* we may expect to see:

a) a higher assimilation rate per unit area, partially offset by lower Rubisco activity per unit area;

b) the same or lower stomatal conductance per unit area;

c) a change in leaf energy balance, that may lead to higher leaf temperature and partially offset the potential reduction in transpiration rate (E).

How these responses of Rubisco, stomata and temperature, with their interacting feedback pathways, may interact to influence water use efficiency (A/E) in a high CO_2 environment, which is also likely to experience change with respect to other variables of radiation, atmospheric temperature and humidity, is problematical. Useful predictions in particular cases may result from the application of appropriate models of leaf processes.

Cowan & Farquhar (1977) introduced a control theory approach to this problem through their dE/dA optimisation hypothesis. More recently, the use of control theory with explicit definition of the known feedback pathways has been developed for the response of E to change in g_s in a variable environment (McNaughton & Jarvis, 1991), and in a parallel approach control theory has been applied to determine the sensitivity of A to change in g_s and Rubisco in a variable environment (Woodrow, Ball & Berry, 1990). What is now needed is a combined approach, embracing all definable feedback pathways, that can be applied to analyse the sensitivity of both E and A to small changes in biochemical, physiological and environmental variables at leaf scale.

At *plant scale* we may expect to see

a) substantial but rather variable increases in plant growth and changes in the phenology of growth processes;

b) a number of specific anatomical and morphological differences;

c) more C allocation to roots leading to larger root systems that may be more effective in water uptake;

d) more rapid leaf growth, larger leaves and larger plants with more leaves that may have higher transpiration rates.

Empirical data suggest that increases in plant size and, particularly, in leaf area may offset any possible advantages to water economy resulting from stomatal closure. There is also some empirical evidence to suggest that the growth enhancing effects of elevated CO_2 may to some extent offset the growth reducing effects of water stress (see review of Eamus & Jarvis, 1989) but other recent experiments have shown similar growth reductions as a result of water stress in both current and elevated CO_2 environments (*e.g.* Conroy *et al.*, 1990).

To understand these growth responses to CO_2, to be able to generalise and to make useful predictions about the consequences for plant water relations in an environment in which we may also expect changes in rainfall and atmospheric humidity, we require a physiological, process-based plant growth model.

This model should include the known feedback pathways linking transpiration, C assimilation and N uptake, handle C and N allocation within the plant and the growth of leaf area and root length. To be useful in the present context, such a model must include explicitly known 'pressure points' on which elevated CO_2 acts, such as Rubisco and stomata, but also other possibile sites of direct action, such as root and leaf growth. More critical experimentation is needed to define these 'pressure points', without which any model will not be able to help us to understand how global change may affect plant growth and water relations.

EFFECTS ON WATER RELATIONS AT STAND SCALE

The extent to which these responses observed on leaves and plants, usually growing singly or in small groups, may be apparent at stand scale requires consideration of the degree of coupling between leaves and atmosphere and occupancy of the available soil volume.

The degree of coupling depends on the kind of canopy the leaf is in, and the kind of vegetation of which the plant is part. Coupling, for this purpose, can be regarded as the degree of similarity between the values of saturation deficit and CO_2 concentration at the leaf surfaces and those in the free airstream above the vegetation. If the two are similar, then the leaves can be considered as well-coupled to the atmosphere: if they are dissimilar with the saturation deficit, for example, at the leaf surface substantially independent of that above, the plants are poorly-coupled to the atmosphere (Jarvis & McNaughton, 1986). Poorly-coupled vegetation is generally short (< 1 m); well-coupled vegetation is generally tall (> 2 m). In poorly-coupled vegetation large reductions in stomatal conductance are necessary to reduce transpiration, and CO_2 assimilation, whereas in well-coupled vegetation transpiration and CO_2 assimilation are very sensitive to small changes in stomatal conductance. Thus a decrease in stomatal conductance resulting from an increase in ambient CO_2 concentration may have very different consequences for transpiration and water use by vegetation, depending on the structure of the vegetation.

Similarly, an increase in leaf area will have little effect on transpiration from poorly-coupled vegetation if net radiation is already mostly absorbed. By contrast, transpiration from well-coupled vegetation is open ended with respect to increase in canopy conductance, which increases *pro rata* with leaf area. Thus changes in leaf area resulting from elevated CO_2 have very different consequences for transpiration, again depending on the structure of the vegetation.

At stand scale, too, larger root systems may have substantial consequences for the rate of removal of soil water if the rooting volume is not already fully exploited. However, in a closed community, with both canopy closure and complete occupancy of the rooting volume (*i.e.* root closure), a tendency to increase the mass and length of roots may have little impact on water use. Thus the structure of the vegetation may again determine the consequences of what looks like a fairly straightforward response at the isolated plant scale.

The maintenance of internal plant C/N ratio may also depend on more complex and subtle effects of CO_2 on leaf quality, and climate change will almost certainly affect mineralisation processes, rates of N turnover and availability of water and N. Thus more complex ecosystem models are needed to evaluate the likely impact of global change on the water relations of vegetation.

Clearly, we cannot safely extrapolate from leaf to stand to ecosystem in a simple way. Effects of elevated CO_2 on leaf and plant process are readily definable and need to be defined to understand likely effects on water relations at the plant scale. However, to go further to stand scale, even if only considering water vapour and CO_2 exchange, requires the use of models that take the degree of coupling into account explicitly and can accommodate the substantial negative feedbacks that act to stabilise the larger scale system (McNaughton & Jarvis, 1991). To go still further to ecosystem scale requires the use of soil-vegetation-atmosphere transfer models that extend vertically to include the entire convective boundary layer.

REFERENCES

Acock, B. & Allen, L.H. (1985) Crop responses to elevated carbon dioxide concentration. In *Direct Effects of Increasing Carbon Dioxide on Vegetation* (eds. B.R. Strain & J.D. Cure), pp 53-97. U.S. Department of Energy, Washington D.C. [DOE/ER - 0238].

Conroy, J.P., Milham, P.J., Mazur, M. & Barlow, P.W.R. (1990) Growth, dry weight partitioning and wood properties of *Pinus radiata* D. Don after 2 years of CO_2 enrichment. *Plant, Cell and Environment,* 13, 329-337.

Cowan, I.R. & Farquhar, G.D. (1977) Stomatal function in relation to leaf metabolism and environment. *Symposium of Society for Experimental Biology,* 31, 471-505.

Eamus, D. & Jarvis, P.G. (1989) Direct effects of increase in the global atmospheric CO_2 concentration on natural and commercial trees and forests. *Advances in Ecological Research,* 19, 1-55.

Hammond, A.L. (1990) (ed.) *World Resources 1990-1991.* Oxford University Press, New York.

Houghton, J.T., Jenkins, G.J. & Ephraums, J.J. (1990) (eds.) *Climate Change. The IPCC Scientific Assessment.* Cambridge University Press, Cambridge.

Houghton, R.A. (1991) Tropical deforestation and atmospheric carbon dioxide. *Climate Change* (in press).

Jarvis, P.G. (1989) Atmospheric carbon dioxide and forests. *Philosophical Transactions of the Royal Society London,* B324, 369-392.

Jarvis, P.G. & McNaughton, K.G. (1986) Stomatal control of transpiration. *Advances in Ecological Research,* 15, 1-29.

Jarvis, P.G. & Morison, J.I.L. (1981) The control of transpiration and photosynthesis by the stomata. In *Stomatal Physiology* (Eds. P.G. Jarvis & T.A. Mansfield), pp 247-279, SEB Seminar Series 8, Cambridge University Press, Cambridge.

McNaughton, K.G. & Jarvis, P.G. (1991) Effects of spatial scale on stomatal control of transpiration. *Agricultural and Forest Meteorology,* 54, 279-302.

Meidner, H. & Mansfield, T.A. (1965) *Physiology of Stomata.* 179 pp. McGraw-Hill, London.

Morison, J.I.L. (1987) Intercellular CO_2 concentration and stomatal response to CO_2. In *Stomatal Function* (Eds. E. Zeiger, G.D. Farquhar & I.R. Cowan), pp 229-251, Stanford University Press, Stanford.

Morison, J.I.L. & Gifford, R.M. (1984) Plant growth and water use with limited water supply in high CO_2 concentrations. I. Leaf area, water use and transpiration. *Australian Journal of Plant Physiology,* 11, 361-374.

Mott, K.A. (1988) Do stomata respond to CO_2 concentrations other than intercellular? *Plant Physiology,* **86**, 200-203.

Mott, K.A. (1990) Sensing of atmospheric CO_2 by plants. *Plant, Cell and Environment,* **13**, 731-737.

Parry, M.L. *et al.* (1991) *The Potential Effects of Climate Change in the United Kingdom.* 123pp, HMSO, London.

Radoglou, K.M. & Jarvis, P.G. (1990) Effects of CO_2 enrichment on four poplar clones. I. Growth and leaf anatomy. *Annals of Botany,* **65**, 617-626.

Raschke, K. (1958) Über den Einfluss der Diffusions widerstände auf die Transpiration und die Temperatur eines Blättes. *Flora,* **146**, 546-578.

Sage, R.F., Sharkey, T.D. & Seemann, J.R. (1988) The *in vivo* response of the ribulose-1,5-bisphosphate carboxylase activation state and the pool sizes of photosynthetic metabolites to elevated CO_2 in *Phaseolus vulgaris* L. *Planta,* **174**, 407-416.

Sage, R.F., Sharkey, T.D. & Seemann, J.R. (1989) Acclimation of photosynthesis to elevated CO_2 in five C_3 species. *Plant Physiology,* **89**, 590-596.

Tans, P.P., Fung, I.Y. & Takahasi, T. (1990) Observational constraints on the global atmospheric CO_2 budget. *Science,* **247**, 1431-1438.

Woodrow, I.E., Ball, J.T. & Berry, J.A. (1990) Control of photosynthetic carbon dioxide fixation by the boundary layer, stomata and ribulose 1,5 bisphosphate carboxylase/oxygenase. *Plant, Cell and Environment,* **13**, 339-347.

Woodrow, I.E. & Berry, J.A. (1988) Enzymatic regulation of photosynthetic carbon dioxide fixations in C_3 plants. *Annual Review of Plant Physiology and Plant Molecular Biology,* **39**, 533-594.

Cavitation. A review: past, present and future

J. A. MILBURN

Department of Botany, University of New England,
Armidale, NSW 2351, Australia.

SUMMARY

A major review on xylem and its vulnerability to cavitation and embolisation has been published recently by Tyree & Sperry (1989). This review is intended to augment their more comprehensive overview adding additional background and presenting the subject from a somewhat different viewpoint. The coverage traces the development of an interest in cavitation as a curiosity, using newly available technology, towards the present day when it has become a major field of international research. This review stresses the way in which information has been gleaned from very different model systems, all of which contibute to our understanding of the mechanisms involved in cavitation, embolisation and its reversal. In seeking beneficial results by application of this knowledge, *e.g.* in forestry or horticulture, it is important to remember that a broad scope has much to commend it.

AUDIO DETECTION OF CAVITATION

According to my notebooks we first detected acoustic signals from plants on 6th February 1963. The sounds were produced on a record-player amplification system and filled our converted-museum laboratory in Aberdeen university. The elation produced was partially on account of the champagne-cork sounds which flooded the laboratory, but also from the fact that earlier work in previous months, based on the removal of embolisation, had predicted they might be found. Sadly, validation of such predictions is rare and not often so easily realised in science! The main avenue of this research was the uptake of water by seriously water-stressed leaves. The patterns of water uptake of leaves via the petioles was quite different from floating leaf discs, which can absorb water via their edges without its having to transverse the main conducting elements within the petioles. Since the effects occurred just above freezing-point it seemed likely that some physical mechanism was snapping the columns of water under stress (cavitation) and the strange uptake patterns were a result of their reconstitution (*i.e.* removal of emboli) when a supply was renewed. This work was published (Milburn; Milburn & Johnson) in 1966.

The idea of listening for cavitation was not itself new, because both Berthelot (1850) and Dixon (1914) had listened to cavitation in sealed glass tubes without the need for electronic amplification. Cooling caused by differential contraction between water and its rigid glass envelope caused a tension which was relieved suddenly: there was a sharp "click" and a cloud of small bubbles was produced. Later, others (Ursprung, 1915; Renner, 1915) detected cavitation microscopically in fern sporangia, and later in xylem conduits (*e.g.* Peirce, 1936; Crafts, 1939), though the latter is neither clear-cut nor obvious, because the emptying of xylem conduits might easily arise from the introduction of air via wounds after tissues have been exposed surgically.

At first we regarded cavitation detection as a curiosity, because it seemed likely that it took place only *in extremis,* such was the confidence, then general, in the cohesion hypothesis. However, it was puzzling that cavitation was often detected *before* the leaves under investigation showed visible signs of wilting. So cavitation, as an explanation for unexpected experimental results, had to be established by the accumulation of circumstantial evidence because there was no alternative method with which to compare it at that time (see Milburn, 1973a, 1973b). Only later, when the newly-devised pressure bomb (Scholander *et al.,* 1965) became generally available did the measurement of xylem sap tensions become an easy and reliable prospect. Results so obtained were not necessarily easy to interpret, because the theory of determining xylem sap tensions using a pressure bomb depends on refilling only the air-filled conduits (*i.e.* vessels, tracheids and fibres) with sap. This sap must be squeezed by compressed gas from living cells adjoining the conduits to displace air introduced into them when the plant tissues, and especially the conduits, were cut. If, however, cavitation had already taken place the conduits would not only be empty, but to a considerable extent embolised by gas also. Such doubts were dispelled, however, by calculating the possible magnitude of the possible error from measurements of conduit volumes and also by making independent checks with different instruments, such as psychrometers. These have supported the general accuracy of the pressure chamber measurements; errors seem generally to be insignificant. I am sceptical about doubts raised recently against the accuracy of the pressure chamber by Balling & Zimmermann (1990).

FACTORS INDUCING CAVITATION

Since antiquity cavitation had been recognised as a major problem when raising water, hence also sap ascent up tall trees; however, as research has progressed we have been forced to reassess this view. Measurements and calculations show that gravitational forces are minor when compared with the difficulty of extracting water from dry or saline soils. The mechanical effects of wind, once thought to be a major problem, now seem to be overcome by the porous, and therefore somewhat elastic and resilient design of the vascular system. Attempts to explain its functioning have been made since Bailey (1916).

Theoretically, based on devices such as the bubble chamber which can be used to track sub-atomic particles by their cavitation trails, one might expect radioactive bombardment to induce cavitation in plants but there is little evidence of this (Milburn, 1973a). On the other hand, any particulate suspension which can block the xylem pit membranes quickly induces cavitation in a transpiring shoot and even sugar solutions, presumably on account of their viscosity, have the same effect (Milburn, 1973b). Many of these effects, which can be induced by wilt inducing pathogens also, can be understood simply enough on the basis of a restriction in water supply; a condition which can be demonstrated directly using excised leaves.

In a somewhat similar manner it is easy to understand how a restriction in water-loss can diminish cavitation; water-loss would normally be restricted by stomatal closure, which can be counteracted by chemicals introduced by pathogens, such as fusicoccin (Turner, 1972). Evaporation can be controlled more directly by enclosing a plant or plant organ in a polythene bag. Recent work has shown this technique can extend the vase-life of cut shoots of *Acacia* very considerably while also reducing their rate of cavitation (Williamson & Milburn, unpublished). Water loss from leaves of plants can be suddenly increased quite naturally by direct sunlight and this seems to be a major cause of cavitation in small herbs and possibly trees also.

Further consideration of xylem anatomy and ultrastructure is appropriate at this point. Though the general anatomy of xylem has been reasonably understood for some time, this knowledge was quite inadequate to allow the numbers of acoustic emissions to be correlated with actual anatomical units. Anatomists had tended to concentrate on the cellular elements from which vessels were constructed as an aid to wood identification and classification. However, the physiological unit of importance was that bounded by ultramicroscopic membranes, so the old and disused term *conduit* was revived (Milburn & Covey-Crump, 1971) to cover the possibility that vessels, tracheids or fibres might all contribute to the cavitation-induced acoustic emissions.

The discovery that cavitation occurred rather readily in plants helped to explain one long-standing problem. Why was the course of xylem sap so frequently intercepted by pit membranes, which had pores so fine that they could filter the fine carbon particles from indian ink and must therefore constitute a significant hydraulic resistance? The answer is now quite obvious: without them, a bubble produced by cavitation would spread progressively and catastrophically throughout the whole system. Pit membranes allow the passage of sap or water but not particles, or more importantly bubbles, to pass; this latter because of surface tension which is exerted at interfaces. Such interfaces are particularly effective in very narrow pores, such as occur in pit membranes. But did the number of conduits, which were of so many different diameters and lengths, coincide with the numbers of acoustic signals? The use of isolated vascular bundles, in comparison with bundles enclosed in petioles (Milburn & McLaughlin, 1974), showed that when enclosed within plant tissue only a fraction of the total number of conduits actually cavitated, because subsequent exposure produced a large additional set of acoustic signals. Apparently, when enclosed only the "weak links" in the "chains" are broken, but exposure breaks all of the links. This work also showed that conduits could be restored with water, so regaining the capacity to cavitate anew many times. It had seemed possible that the walls or pits might be seriously damaged by cavitation. It now seems that permanent damage of this kind is rather infrequent because, after restoration with water, vascular bundles produce very similar numbers of clicks to those produced originally. The number of clicks was quite small, a few hundred per bundle, which gave rather irregular curves when click frequency was plotted against time.

About 1975 it was becoming apparent that the detailed study of cavitation might actually be useful and not merely a scientific curiosity. If different plant species cavitated at different sap tensions, reflecting adaptations to climatic changes and their ecological situations which require different performance criteria for survival, perhaps these differences could be exploited and measured to distinguish between the more or less cavitation-vulnerable plants. It might also be possible to produce more stress-resistant crops. At the same time it was becoming clear that small herbs and large trees cavitated generally within similar ranges of water potential (-0.5 to - 2.5 MPa). Considerable precision would be needed to distinguish between clones within these ranges. The number of signals detected was insufficient for very high resolution and cavitation: water-stress curves were quite variable between species, individual plants and even different plant organs, such as leaves.

Research was also directed towards exploitation of the favourable characteristics of piezoelectric crystals as vibration detectors. These can generate a high electrical output from small momentum changes which lead to studies of higher frequencies in the mid to late 1970s. This trend towards monitoring higher frequencies was abandoned when it was found that minute specimens of wood when subjected to drying on a piezoelectric acoustic probe gave vastly more acoustic signals than could be related to functional conduits seen in the same specimen under microscopic examination (Milburn & Walding, unpublished). It seemed that possibly signals were being detected from non-conducting cells, or even parts of cells. Since these were unlikely to be related to vascular conduction they were unlikely to be useful when studying the deleterious effects of cavitation. As a result, this approach was abandoned and the major focus was redirected towards measuring the sap tensions which induced cavitation in different species. It had become clear that the xylem was much more vulnerable than previously supposed, a view emphasised by a chapter in my monograph (Milburn, 1979) "Xylem - the vulnerable pipeline".

As indicated earlier, small herbs appeared to cavitate at surprisingly severe tensions in comparison with trees, despite the much greater distances water has to be drawn against gravity within trees. Evidence from herbs was not only based on cavitation data (Milburn & McLaughlin, 1974) but also from some very interesting calculations, based on Poiseuille's equation for flow though conduits, made by Passioura (1972). These indicated that severe tensions *must* develop in small transpiring wheat seedlings. However, there could be no doubt that trees (*e.g. Fraxinus*) which were able to survive water-stress and drought, were also able to sustain very high tensions before cavitation became serious (see Zimmermann & Milburn, 1982). In the course of this research, and also in subsequent studies, many species have been added to illustrate the range (Crombie, Milburn & Hipkins, 1985). These include shrubs, trees and interestingly a gigantic herb - the banana plant. This plant apparently cavitates readily at quite low sap tensions (-0.5 to -1.0 MPa, Milburn & Ritman, unpublished): however, normally it lives with even lower xylem sap tensions in a remarkably hydrated condition. Even at midday, banana xylem sap tensions were seldom more severe than 0.2 MPa and even under prolonged drought they did not exceed 0.4 MPa

(Milburn, Kallarackal & Baker, 1990; Kallarackal, Milburn & Baker, 1990).

ULTRASONIC AND AUDIO TECHNIQUES

In the early 1980s ultrasonic techniques were applied to the study of cavitation in the range of 0.2-2.0 MHz, which is much higher than the range found previously through experience to be most useful *i.e.* 500-20,000 Hz in the audio range. It was immediately apparent that the results were quite different because far more signals were detected than in the audio range (thousands and tens of thousands from small specimens of wood). Furthermore, the events appeared to be relatively free from interference, which is troublesome in the audio range.There was a sudden upsurge in interest in cavitation detection from this period (Tyree & Dixon, 1983; Tyree *et al.,* 1984; Sandford & Grace, 1985; Jones & Pena, 1986; Pena & Grace, 1986; Salleo & Lo Gullo, 1986).

So far as could be determined the incidence of cavitation determined ultrasonically corresponded well with the results of work in the audio ranges. However, it seemed impossible that the acoustic events could correspond to the *same* events because of the disparity in numbers (Ritman & Milburn, 1988). Recently this problem has been examined using a double probe (Ritman & Milburn, 1990) which detected audio acoustic emissions (AAE) simultaneously with ultrasonic acoustic emissions (UAE). As a result, we can be reasonably sure that the events are *roughly* correlated. However, AAE and UAE are not always detectable from the same tissues; while some appear to produce AAE, other tissues produce UAE, while yet others produce both.

Generally AAE tend to be produced *slightly earlier* than UAE *i.e.* at more moderate sap tensions. It does not seem that the two are manifestations of the same event within the plant; this raises the vexed question of their nature and origin.

Another feature is quite apparent also. While AAE can be detected over quite long distances (ca. 1 m) within the plant (Milburn, 1973b; Ritman & Milburn, 1990), UAE are of quite local origin (ca. 20 mm from the probe). It seems therefore that both techniques have useful features which can be used to complement one another. Apparently a detection system is liable to influences from the tissue in which cavitation occurs, because tissues can act as acoustic filters, damping out some vibrations selectively.

This work has been extended recently to investigate the relationship of conduit length with the frequency of emission. The general idea is that the larger the conduit the lower the frequency emitted because it will function acoustically rather like an organ pipe or a drum. Thus it might be anticipated that whilst vessels, being the largest units, would generate low frequency sounds (AAE), tracheids and fibres might contribute the higher frequency vibrations (UAE). There is some evidence (Ritman & Milburn, 1991) that this is broadly true. However, by studying vessel-less species it has been possible to say with certainty that vessels are *not* essential for the production of AAE because they are produced by species with tracheids and fibres only.

Some have questioned if the size of conduits is relevant when measuring the frequency of AAE and UAE (Tyree & Sperry, 1989). In my view this must remain a moot point until we know exactly how the vibrations are produced. Unquestionably the vibrations are produced by the tissue and there can be little doubt the conduits themselves are responsible. However, the strains on the walls in the intact situation have not been measured and furthermore the movements of the walls are restrained and damped by surrounding tissues which include both living, turgid cells and dead fibres. It seems most likely that ultrasonic methods include tissue fracture signals in large numbers.

One approach to the elucidation of these questions has been to use much less complex systems such as fern sporangia. In these, the tensions inducing cavitation are enormous (ca. 20 to 35 MPa). The great advantage is that they can be observed microscopically during cavitation without the need for surgery, which might damage conduits during their extraction.

Techniques have been developed (Ritman & Milburn, 1990) for the observation of individual sporangia during cavitation while simultaneously monitoring for AAE and UAE. These studies have shown that each sporangium produces up to 3 or 4 AAE as it reaches the stage when cavitation is taking place. These seem to be associated with major movements such as the wall tearing and release of the sporangium and coincide with changes in position. UAE are more numerous and about 20-24 are produced, which coincide roughly with the number of annular cells. These have thick walls and do not produce visible changes in position as cavitation bubbles are produced. Broadly then, these studies indicate that larger movements coincide with AAE, but UAE are produced by small thick-walled cells which may well correspond with cavitation or damage within fibres or even thickened parenchymatous cells in xylem tissues (Ritman & Milburn, 1991).

CAVITATION AND HYDRAULIC CONDUCTION

When cavitation occurs within a conduit, a bubble is produced at near-vacuum pressure; only a minute quantity of gas, with a larger quantity of water vapour is present. If this tension is relieved quickly, by supplying water, the bubble so formed shrinks and disappears rapidly. If the bubble is held for a period of time however, gases diffuse through the tissues and invade the bubble so raising its internal gas pressure towards atmospheric pressure (0.1 MPa). Such a bubble is now a gas-filled embolism. Both kinds of bubble will prevent conduction via a conduit; the main difference between them lies in the ease of removal. Gas-filled emboli are much more difficult to remove than evacuated bubbles. However, pressurisation, vacuum injection, or the use of degassed (previously boiled or evacuated) water will all promote their dissolution.

In recent years several studies have shown that cavitation does reduce hydraulic conduction along the lines predicted (Dixon, Grace & Tyree, 1984; Crombie, Milburn & Hipkins, 1985; Sperry, Donnelly & Tyree, 1987). This work has been performed mainly using UAE combined with techniques measuring the flow of water through pieces of woody tissue. Much more work of this kind needs to be performed on a range of species before we can be certain how the stage at which cavitation seems to occur most frequently is related to the greatest loss in hydraulic conductivity, or in the words of Tyree & Sperry (1989), "hydraulic sufficiency".

THE MECHANISM OF CAVITATION

The simplest way to visualise cavitation is to observe the Berthelot tube experiment. The walls of the tube are pulled inwards and then vibrate when cavitation occurs producing AAE. When plant conduits cavitate they, having no comparable mechanism of differential expansion, must lose water through their minutely porous walls. Air cannot enter, providing the pores are wet and sealed by surface-tension via a film of water. Of course if the strain becomes sufficiently severe, gas (mainly air) could be drawn through a pore (see Pickard, 1981; Zimmermann, 1983). A single tiny bubble inside the conduit would instantly produce total disruption, because once it enters it would immediately expand until the strain within the encircling walls has been relieved. The second phase, which can be seen with appropriate microscopic preparations (a video recording can be used to illustrate this point), is when the bubble continues to enlarge, relatively slowly, until it completely fills the conduit with gas at near vacuum pressure.

It is highly unlikely that at the relatively modest tensions exerted in our experiments the hydrogen bonding between water molecules can be overcome: such pressures are several magnitudes greater than those generated in plants. But it is possible that the initial small bubble is not actually drawn through the porous walls but expands from a tiny initial bubble, or defect of some kind, held inside the conduit, perhaps inside a piece of debris attached to the inner walls. Evidence points to the possibility of both mechanisms.

On one hand Crombie, Hipkins & Milburn (1985) discovered that the addition of surface-active chemicals could precipitate cavitation at reduced tensions, apparently through lowering the surface tension of films which normally kept the conduits "air-tight". Similar findings were made by injecting pressurised air into xylem (see also Sperry & Tyree, 1988). According to these results, air enters conduits from the outside.

Proving the alternative mechanism in conduits is a more difficult task but it is still in my view a real possibility. We know that cavitation can be induced in fern sporangia and fungal ascospores while these are *completely submerged* in osmotic fluids, without access to atmospheric air (Milburn, 1970). Indeed it would appear in the latter instance that cavitation is induced within the living cytoplasm of the cell. Unfortunately vascular conduits have not as yet been subjected to this kind of experiment. Their walls are porous, as distinct from semi-permeable, and the only means of generating tensions has been through evaporative loss of water into an adjoining gas phase, which prevents unambiguous conclusions being drawn.

THE INTRODUCTION OF EMBOLI INTO CONDUITS

Following cavitation the permanentising of the blockage through the evolution of gas (embolisation) seems to be quite a rapid process. The simplest way to see this is by using fungal spores under the microscope. If a water supply is rapidly renewed following cavitation the gas bubble takes little time to disappear. Delay in the renewal of a water supply slows down the process considerably (Milburn, 1970). In theory the same thing happens within a vascular conduit; but the path of diffusion may not always be so short as inferred in a recent review (Tyree & Sperry, 1989). In my view embolisation in tightly-knit water-filled vascular tissues may take considerably longer than might be supposed for a single conduit surrounded in air. It is not easy to design experiments to demonstrate this effect precisely, but it will be noted that the research which lead to the detection of cavitation (Milburn, 1966) was based on reversal of embolisation, and this gives a clue to the time scales involved. I suspect the process may, in some instances, take several hours to complete, or even longer. Great strides have been taken in the study of embolisation through changes in hydraulic conductance (cf. Newbanks, Bosch & Zimmermann, 1983; Sperry, Donnelly & Tyree, 1987; Tyree & Sperry, 1988; Sperry & Tyree, 1988). These studies provide very powerful support for the present understanding of xylem sap transport in general. They also address a key issue, viz. the availability of transported water to plants in general. More work of this kind is needed, and it is especially important to elucidate the role of living cells which appear both to precipitate embolisation in some instances (Coutts, 1977), and maintain the vascular xylem as a functional unit in general. Evidence is accumulating that embolisation can be reversed: it remains to be seen if rainwater plays an important role. This has been indicated by some of our earlier work and also the work of Katz *et al.* (1989), which later used the same techniques.

Another interesting problem hinges on the effect of low temperatures on the induction of cavitation. According to Briggs (1950) water, hence by analogy xylem sap, becomes exceptionally vulnerable to cavitation at low temperatures in the vicinity of 0-5 °C. However, experience would suggest that plants do not become vulnerable in this way until temperatures fall *below zero* when sap may freeze. In addition, experiments on vascular plants and fern sporangia (Ritman, 1988) and also fungal spores (Milburn, unpublished), indicate that sap is, if anything, *less* vulnerable to cavitation at temperatures just above freezing point, in opposition to Briggs results.

A similar series of very interesting experiments concerns embolisation in the genus *Acer*. Many workers have tried to understand the mechanism of maple sap flow, on which the sugar maple industry is based (*e.g.* Marvin, 1958). Long ago it had been observed that the xylem fibres tended to be gas-filled at time maple sap flow occurred (Wiegand, 1906). This system was later proposed by Milburn & O'Malley (1984) to provide a pneumatic "spring" causing the pressurisation of sap during freezing and thawing cycles; hence exudation when the tree has thawed and been wounded via bore holes. In this case the embolisation occurred *selectively* within the wood fibres but *not* within the vessels themselves, showing that the tree had some capacity to control embolisation within tissues, allowing it only in fibres where embolisation would cause least disruption to sap conduction. The work cited showed that freezing operated on embolised *Acer pseudoplatanus* fibres; later, Tyree (1983) showed that the same mechanism operated in *Acer saccharum*, the sugar maple itself, and changes in embolisation have been extensively documented recently by Sperry, Donnelly, & Tyree (1988).

ACKNOWLEDGEMENTS

I express my gratitude to the Organisers of the Vallombrosa International Workshop for their support, the E.E.C. sponsors of the meeting, and to participants who through their enthusiasm and criticism made the project so interesting.

REFERENCES

Bailey, I.W. (1916) The structure of the bordered pits of conifers and its bearing upon the tension hypothesis of the ascent of sap in plants. *Botanical Gazette*, **62**,133-142.

Balling, A. & Zimmermann, U. (1990) Comparative measurements of the xylem pressure of *Nicotiana* plants by means of the pressure bomb and pressure probe. *Planta*, **182**, 325-338.

Berthelot, M. (1850) Sur quelques phénomènes de dilatation forcée des liquides. *Annales de Chimie et Physique*, 3e Sèr. **30**, 232-237.

Briggs, L.J. (1950) Limiting negative pressure of water. *Journal of Applied Physics*, **21**, 721-723.

Coutts, M.P. (1977) The formation of dry zones in the sapwood of conifers. II. The role of living cells in the release of water. *European Journal of Forest Pathology*, **7**, 2-6.

Crafts, A.S. (1939) Solute transport in plants. *Science*, **90**, 337-338

Crombie, D.S., Hipkins, M.F. & Milburn, J.A. (1985) Gas penetration of pit membranes in the xylem of *Rhododendron* as the cause of acoustically detectable sap cavitation. *Australian Journal of Plant Physiology*, **12**, 445-453.

Crombie, D.S., Milburn, J.A. & Hipkins, M.F. (1985) Maximum sustainable xylem sap tensions in *Rhododendron* and other species. *Planta*, **163**, 27-33.

Dixon, H.H. (1914) *Transpiration and the Ascent of Sap in Plants*. MacMillan, London.

Dixon, M.A., Grace, J. & Tyree, M.T. (1984) Concurrent measurements of stem density, leaf and stem water potential, stomatal conductance and cavitation on a shoot of *Thuya occidentalis* L. *Plant, Cell & Environment*, **7**, 615-618.

Kallarackal, J., Milburn, J.A. & Baker, D.A. (1990) Water relations of the banana. III Effects of controlled water stress on water potential, transpiration, photosynthesis and leaf growth. *Australian Journal of Plant Physiology*, **17**, 79-90.

Katz, C., Oren, R., Schultze, E.D. & Milburn, J.A. (1989) Uptake of water and solutes through the aerial organs of *Picea abies* (L) Karst. *Trees,* **5**, 1-5.

Jones, H.G. & Pena, J. (1986) Relationship between water stress and ultrasound emission in apple *(Malus domestica* Borkh.). *Journal of Experimental Botany,* **37**, 1245-1254.

Marvin, J.W. (1958) The physiology of maple sap flow. In *The Physiology of Forest Trees* (ed. K.V. Thimann, W.B. Critchfield & M.H. Zimmermann). The Ronald Press Co., New York.

Milburn, J.A. (1966) The conduction of sap. I. Water conduction and cavitation in water stressed leaves. *Planta,* **65**, 34-42

Milburn, J.A. (1970) Cavitation and osmotic potentials of *Sordaria* ascospores. *New Phytologist,* **69**, 133-141.

Milburn, J.A. (1973a) Cavitation in *Ricinus* leaves by acoustic detection: induction in excised leaves by various factors. *Planta,* **110**, 253-265.

Milburn, J.A. (1973b) Cavitation studies on whole *Ricinus* plants by acoustic detection. *Planta,* **112**, 333-342.

Milburn, J.A. (1979) *Water Flow in Plants.* Longman, London New York.

Milburn, J.A. & Covey-Crump, P.A.K. (1971) A simple method for the determination of conduit length and distribution in stems. *New Phytologist,* **70**, 427-434.

Milburn, J.A., Kallarackal, J. & Baker, D.A. (1990) Water relations on the banana. I. Predicting the water relations of the field-grown banana using the exuding latex. *Australian Journal of Plant Physiology,* **17**, 57-68.

Milburn, J.A. & Johnson, R.P.C. (1966) The conduction of sap. II. Detection of vibrations produced by sap cavitation in *Ricinus* xylem. *Planta,* **66**, 43-52.

Milburn, J.A. & McLaughlin, M.E. (1974) Studies of cavitation in isolated vascular bundles and whole leaves of *Plantago major* L. *New Phytologist,* **73**, 861-871.

Milburn, J.A. & O'Malley, P.E. (1984) Freeze-induced sap absorption in *Acer Pseudoplatanus*: a possible mechanism. *Canadian Journal of Botany,* **61**, 3100-3106.

Newbanks, D., Bosch, A. & Zimmermann, M.H. (1983) Evidence for xylem dysfunction by embolization in Dutch elm disease. *Phytopathology,* **73**, 1060-1063.

Passioura, J.B. (1972) The effect of root geometry on the yield of wheat growing on stored water. *Australian Journal of Agricultural Research,* **23**, 745-752.

Peña, J. & Grace, J. (1986) Water relations and ultrasound emissions of *Pinus sylvestris* L. before, during and after a period of water stress. *New Phytologist,* **103**, 515-524.

Peirce, G.J. (1936) The state of water in ducts and tracheids. *Plant Physiology,* **11**, 623-628.

Pickard, W.F. (1981) The ascent of sap in plants. *Progress Biophysical Molecular Biology,* **37**, 181-229.

Renner, O. (1915) Theoretisches und Experimentelles zur Kohèsionetheories der Wasserbewegung. *Jahrbuecher für Wissenschaftliche Botanich.,* **56**, 617-667.

Ritman, K.T. (1988) *Plant Water Relations: Investigations into Acoutic Emission and Transmission.* Ph.D. Thesis submitted to the University of New England, Australia.

Ritman, K.T. & Milburn, J.A. (1988) Acoustic emissions from plants: ultrasonic and audible compared. *Journal of Experimental Botany*, **39**, 1237-1248.

Ritman, K.T. & Milburn, J.A. (1990) The acoustic detection of cavitation in fern sporangia. *Journal of Experimental Botany*, **41**, 1157-1160.

Ritman, K.T. & Milburn, J.A. (1991) Monitoring of ultrasonic and audible emissions from plants with or without vessels. *Journal of Experimental Botany*, **42**, 123-130.

Salleo, S. & Lo Gullo, M.A. (1986) Xylem cavitation in nodes and internodes of whole *Chorisia insignis* H.B. et K. plants subjected to water stress: relations between xylem conduit size and cavitation. *Annals of Botany*, **58**, 431-441.

Sandford, A.P. & Grace, J. (1985) The measurement and interpretation of ultrasound from woody stems. *Journal of Experimental Botany*, **36**, 298-311.

Scholander, P.F., Hammel, H.T., Bradstreet, E.A. & Hemmingsen, E.A. (1965) Sap pressure in vascular plants. *Science*, **148**, 339-346.

Sperry, J.S., Donnelly, J.R. & Tyree, M.T. (1987) A method for measuring hydraulic conductivity and embolism in xylem. *Plant, Cell & Environment*, **11**, 35-40.

Sperry, J.S., Donnelly, J.R. & Tyree, M.T. (1988) Seasonal occurrence of xylem embolism in sugar maple *(Acer saccharum)*. *American Journal of Botany*, **75**, 1212-1218.

Sperry, J.S. & Tyree, M.T. (1988) Mechanism of water stress-induced xylem embolism. *Plant Physiology*, **88**, 581-587.

Turner, N.C. (1972) Fusicoccin: a phytotoxin that opens stomata. In *Phytotoxins in Plant Diseases* (ed. R.K.S. Wood, A. Ballio & A. Graniti), pp. 399-401. Academic Press, New York.

Tyree, M.T. (1983) Maple sap uptake, exudation and pressure changes correlated with freezing exotherms and thawing endotherms. *Plant Physiology*, **73**, 277-285.

Tyree, M.T. & Dixon, M. (1983) Cavitation events in *Thuja occidentalis* L. Ultrasonic acoustic emissions from the sapwood can be measured. *Plant Physiology*, **72**, 1094-1099.

Tyree, M.T., Dixon, M.A., Tyree, E.L. & Johnson, R. (1984) Ultrasonic acoustic emissions from the sapwood of cedar and hemlock: an examination of three hypotheses concerning cavitation. *Plant Physiology*, **75**, 988-992.

Tyree, M.T.& Sperry, J.A. (1988) Do woody plants operate near the point of catastrophic xylem dysfunction caused by dynamic water stress? Answers from a model. *Plant Physiology*, **88**, 574-580.

Tyree, M.T. & Sperry, J.S. (1989) Vulnerability of xylem to cavitation and embolism. *Annal Review Plant Physiology Molecular Biology*, **40**, 19-38.

Ursprung, A. (1915) Über die Kohèsion des Wassers im Farnanulus. *Berichte der Deutschen Botanischen Gesellschaft*, **33**, 153-162.

Wiegand, K.M. (1906) Pressure and flow of sap in the maple. *American Naturalist*, **40**, 409-453.

Zimmermann, M.H. (1983) *Xylem Structure and the Ascent of Sap*. Springer-Verlag, Berlin.

Zimmermann, M.H. & Milburn, J.A. (1982) Transport and storage of water. In *Encyclopaedia of Plant Physiology*, New Series (ed. O.L. Lange, P.S. Nobel, C.B. Osmond & H. Ziegler) **12 B**, 135-151. Springer-Verlag, Berlin.

Effect of cavitation on the status of water in plants

J.J. OERTLI

*Institute of Plant Sciences, ETH, 8092 Zürich,
Switzerland.*

SUMMARY

Negative turgor pressures must develop not only in the xylem but also in leaf cells
when the moisture stress reaches a certain level. Since the pressure external to plant
organs is atmospheric, cells containing liquids at negative pressures are exposed to a
compressive stress and will collapse if the stress exceeds a critical limit. Water at
negative turgor pressures is metastable and cavities filled with air should be formed
either through bubble formation or through air entry through pores. If the pressure in
the liquid drops below the vapour pressure of water, the liquid becomes unstable with
respect to the formation of bubbles that are filled with water vapour. Although the
water is metastable, a change to the stable gaseous phase is hindered by a high
"activation energy" due to the surface work required to create a bubble of a critical
size. Cavitation should increase the water potential in plant tissues. In the xylem, this
effect is overshadowed by a decrease in xylem conductivity resulting in a loss of foliar
water potential. Deviations from the usual pressure-volume curves are interpreted to
reflect gains in water potential.

INTRODUCTION

The cohesion theory as originally developed by Böhm in the 1880s
(Böhm, 1893) and subsequently expanded by Askenasy (1895) and Dixon
& Joly (1895) to explain the rise of sap in tall plants had to assume a
metastable state of water that was difficult to reproduce in laboratory
experiments. It was, therefore, not surprising that the theory was
criticized, in particular by those scientists who had a thorough
understanding of physics, and it was even insinuated that Böhm might be
suffering from hallucinations. Today, the cohesion theory finds little
opposition; only occasionally are there expressions of doubts regarding
the existence of metastable states of water in the plant system. Physicists
have even made use of liquids under tension in the successful application
of the bubble chamber although water is apparently unsuitable. The

acceptance of negative turgor pressures, however, is restricted to the xylem and the general belief is that it does not occur in living cells (Lange & Lösch, 1980). Turgor pressures, calculated as the difference between total water potential and its osmotic component, sometimes yield negative values. Tyree (1976), however, demonstrated that contamination of cellular extracts by apoplastic water made such calculations highly questionable and since, beyond the turgor loss point, pressure-volume curves yield straight-line relations between tissue water content (solute concentration) and the reciprocal of the water potential, negative turgor pressures are no longer considered to be scientific reality. However, based on the generally accepted cell model, negative turgor pressures must necessarily develop with severe moisture stress (Oertli, 1989a). Its magnitude is small in mesophytic tissues and could hardly be detected in pressure bomb experiments; in sclerophyllous tissues, however, substantial negative turgor pressures must be expected (Oertli, Lips & Agami, 1990).

This paper is concerned with the development of negative turgor pressure and its consequences for plant tissues. It will be shown that a metastability of water does not necessarily require an instantaneous change in phase. Cavitation, one of the results of negative turgor pressure, will lead to an overall gain in water potential.

The following conventions are adopted: turgor pressure is equal to the absolute pressure minus the external pressure, the latter is usually assumed to be 0.1 MPa (1 bar). Thus, the turgor pressure represents the difference between the actual pressure and that of the standard state of water. Moreover, I shall use pressure units for expressing the water potential and assume that the turgor pressure and the osmolalities are uniquely related to the respective potential components. The last two conventions, although thermodynamically incorrect (Oertli, 1989b), are in general use.

Occasionally cavities are defined as vapour filled spaces in a liquid whereas the term "bubble" (e.g. soap bubble) is restricted to spherical volumes bounded by a membrane and having gaseous phases on both sides and thus having two surfaces to be considered. Here, there is no need to make such a distinction and I shall use the terms interchangeably.

DEVELOPMENT OF NEGATIVE TURGOR PRESSURES

Water in cell walls and within cells is always close to equilibrium (Oertli, 1989) so that for all practical purposes one has (Fig. 1):

$$\Psi_{xylem} = \Psi_{wall} = \Psi_{cell} \qquad \text{Equation 1}$$

where Ψ is the water potential. Except for some specific cases, the osmolalities in the xylem and in adjacent cell walls are negligible. Therefore, in terms of pressure (P) and osmotic (π) components, one has:

$$\Psi_{P\ xylem} = \Psi_{P\ wall} = \Psi_{P\ cell} + \Psi_{\pi\ cell} \qquad \text{Equation 2}$$

Since the osmotic potential component in the cell sap is always negative, ($\Psi_\pi < 0$) it follows:

$$\Psi_{P\ xylem} = \Psi_{P\ wall} < \Psi_{P\ cell} \qquad \text{Equation 3}$$

or

$$P_{xylem} = P_{wall} < P_{cell} \qquad \text{Equation 4}$$

The latter equation states that the pressure within the cell is always larger than the pressure in the cell wall sap and in the adjacent xylem vessel. The plasmamembrane is always pressed against the cell wall and plasmolysis is impossible. As a consequence, if the water potential drops further after the cell has reached the turgor loss point, the decrease in the water potential within the cell will be absorbed not only by the osmotic component (as is the case with plasmolysis) but also by a pressure component, *i.e.* the turgor pressure must necessarily become negative, the extent of which depends on mechanical properties of the cell wall.

CONSEQUENCES OF NEGATIVE TURGOR PRESSURE

(a) *Cell collapse*: Since a plant leaf is exposed to atmospheric pressure, the negative pressure that must occur during severe moisture stress will expose the cell to a compressive stress which, depending on the strength of the walls, will cause the cells to collapse through buckling and sagging of walls (Fig. 2). Oertli *et al.* (1990) have shown that mesophytic tissues show very little resistance to collapse and a negative turgor pressure of a few kPa to several 10 kPa is sufficient to cause collapse. The cell thus behaves as if no wall were present and this explains the more or less linear behaviour often observed beyond the turgor loss point in PV curves. It must be emphasized though, that this linear behaviour is only possible because a small negative turgor pressure has developed. Cell collapse is probably a frequent phenomenon in drought-stressed tissues. It is, to some extent, reversible.

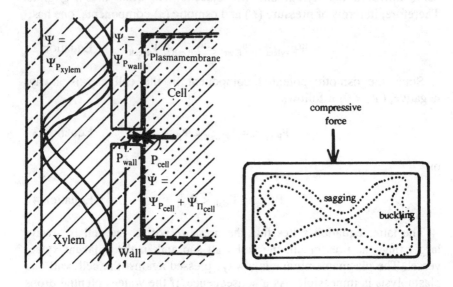

FIG. 1. (left) Near equilibrium of water in the xylem, adjacent cell wall and cell.
FIG. 2. (right) Collapse of cells due to negative turgor pressures.

(b) *Cavitation*: A cell solution that is in equilibrium with the surrounding atmosphere becomes metastable as soon as the internal pressure drops below atmospheric.

Nitrogen and oxygen that are dissolved in the liquid and are in equilibrium with the surrounding atmosphere should form air bubbles at atmospheric pressure within the liquid. As will be discussed below, I do not consider this an important mechanism of cavitation.

External air could also enter the cell through pores in the wall. The air entry value can be derived from the capillary rise equation. The pressure difference across the meniscus in the capillary (Fig. 3) is:

$$\Delta P = (2\gamma \cos\alpha) / r \qquad \text{Equation 5}$$

where γ is the surface tension, r the radius of the capillary and α the contact angle. Since atmospheric pressures above the liquid in the capillary and above the free liquid are equal, ΔP is also the pressure difference in the capillary liquid between the level of the free water surface and the meniscus. In order to bring the water level in the capillary down to that of the free water, an excess pressure of ΔP has to be applied above the meniscus. Obviously ΔP is the limiting value for air entry. The same pressure difference determines the air entry into a pore. In this case, the liquid pressure is lower and the external pressure is atmospheric.

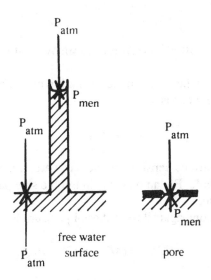

FIG. 3. Relation between capillary rise and air entry value. (P_{men} : pressure below the meniscus.)

For a wetting angle of 0° one obtains the following air entry values:

pit openings 0.1 to 1 μm   air entry 3 - 0.3 MPa
cell wall pores 1 to 10 nm: air entry 300 to 30 MPa
channels in membranes 0.1 nm: air entry 3000 MPa

Considering that the most narrow neck in each pore is critical for air entry, it appears that air entry is only of importance for pit openings. This conclusion is valid even though it is not known to what extent the simple capillary rise equation is valid when pore diameters approach the size of water molecules (Skaar, 1988). Bordered pits are fascinating mechanisms which prevent the spread of cavitation from one xylem compartment into a neighbouring one. Incidentally, the air entry value of bordered pits is not given by the pit size but by the size of the largest holes in the fine meshwork supporting the torus.

A second metastability is reached when the hydrostatic pressure drops below the vapour pressure of water. Spontaneous bubbles should form in the water, *i.e.* the water should boil. Since the cohesion theory proposes negative turgor pressures in the order of megapascals, a change in phase from liquid water to gaseous bubbles (boiling) should theoretically occur in the plant system.

HOW "UNSTABLE" IS METASTABLE

Two work terms are involved in the formation of gas bubbles.
(a) The surface work is:

$$\gamma A = 4\pi r^2 \gamma \qquad \text{Equation 6}$$

where γ is the surface tension at the air/water interface and A is the surface of the bubble. For an increase in bubble size positive surface work has to be performed.
(b) The work done against the external pressure is:

$$V\Delta P = (4/3)\pi r^3 (P_L - P_{Va}) \qquad \text{Equation 7}$$

where V is the volume of the bubble, P_L is the (external) pressure in the liquid and P_{Va} is the vapour pressure within the bubble. If $P_L - P_{Va} < 0$ then the work done is negative, *i.e.* work is gained. These work terms and their sum are shown in Fig. 4 for two levels of moisture stress. Although the water is metastable, work is required to initiate a bubble and only after the bubble has reached a certain critical size where the energy of formation is maximum will the change in phase take place spontaneously. There is thus a barrier to a change in phase similar to the activation energy that prevents the decomposition of most biological molecules such as sucrose which should spontaneously change to CO_2 and water. With increasing stress this activation energy for bubble formation decreases and a change in phase becomes more likely.

There is considerable evidence that liquid water contains flickering clusters of water molecules. As a consequence, the energy relations in Fig. 4 may not apply to extremely small bubbles because conditions could differ between and within clusters. It may be that extremely small bubbles are often formed in metastable water which will collapse again due to unfavourable energy relations. Only when the bubble has reached a critical size will it grow spontaneously. The likelihood of such an event may be small under usual moisture stresses but it needs to occur only once to cause cavitation. The probability of cavitation occurring increases with moisture stress because a smaller sized bubble can already induce cavitation.

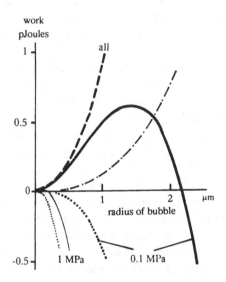

FIG. 4. Surface (dashed), pressure (dotted) and total work (solid lines) required for bubble formation at water stresses of 0.1 (thick lines) and 1 MPa (thin lines). The dash-dotted line represents the latent heat of evaporation.

Since, under otherwise comparable conditions, the probability of cavitation by spontaneous bubble formation increases with the volume of a compartment, one would expect large vessels or cells to cavitate more readily. There is some evidence that this is the case. I am aware though that there may be other explanations for such a phenomenon.

So far I have assumed isothermal conditions. Fig. 4 shows the latent heat of evaporation for bubble formation. This heat energy is supplied by the liquid water and the evaporating surface should become cooler resulting in a lower vapour pressure but the effect should be negligible in the case of cavitation.

The vapour pressure above a curved liquid-air interface differs from a flat surface. This relation is shown by the Kelvin equation:

$$\ln (p_o / p_{bubble}) = 2\gamma V_m / rRT \qquad \text{Equation 8}$$

where p refers to vapour pressures, p_o being that above a flat surface, V_m is the molar volume of water, R is the gas constant and T is the absolute temperature. The reduction of vapour pressure in a bubble of $r=1$ μm is 0.1%. It increases as the bubble becomes smaller and is 10% for a bubble with a radius of 10 nm. These changes in vapour pressure are too small to be of significance.

Except for a very mild moisture stress, this conclusion can be generalized to all aspects affecting the vapour pressure of water because the vapour pressure itself is negligible compared with the magnitude of the absolute value of the negative pressure in the liquid phase.

As a water stress develops, a first instability allows air bubbles to form. Only as the stress becomes more severe do bubbles of water vapour become thermodynamically possible. Nevertheless, I am of the opinion that the formation of air bubbles is an unlikely cause of cavitation: I have calculated the outer radius of a shell that contains the amount of dissolved gases or liquid water that is necessary to fill a bubble (the inner radius is that of the bubble). In the case of a bubble being filled with atmospheric gases, it is 3.6 times the radius of the bubble (Fig. 5), whereas for a water vapour bubble it is only 1.003 times the radius of the bubble, *i.e.* in the latter case the shell is so thin that it cannot be shown in the drawing. Thus, there is a substantial diffusion path for dissolved gases, but not for water vapour, to move to the bubble site. Moreover, experimental evidence shows that cavitation occurs only when the moisture stress is substantial (negative turgor pressures in the order of MPa, Milburn, in this volume, Tyree & Sperry, 1989). At this level of stress, the fact that the instability of dissolved gases appears first should be inconsequential.

However, once cavitation has occurred through bubble formation with water vapour, atmospheric gases will diffuse into the bubble until equilibrium is reached when the internal bubble pressure is nearly atmospheric. A reversal of cavitation is now rendered more difficult.

Kuroda (1989) recently suggested that cavitation in a pine tree that had been infected by nematodes was due to vapours of terpenoids.

A question arises: why is it that we are daily able to boil water without any difficulties? The reason is the presence of nuclei. For example, small gas bubbles may be trapped in dents of the vessel wall. Upon heating to the boiling point, these nuclei grow to large bubbles that escape to the surface with only a small bubble remaining in the dent from which another bubble can grow. It is a well known practice to use boiling stones or to immerse capillaries sealed on one side in sulfuric acid in order to prevent overheating.

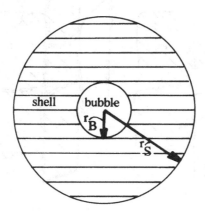

FIG. 5. Size of a shell of liquid water surrounding a bubble and containing dissolved air necessary to fill the bubble at room temperature.

It is speculative as to what extent sites for nucleation exist in the plant system. For example, the cytoplasm is structured and may contain suitable surface features on which bubbles could form more readily than in a free liquid.

In summary, air entry through pores in pits is probably the most likely cause of cavitation (compare Sperry & Tyree, 1988), whereas air entry through cell walls or spontaneous bubble formation within the cellular liquid are extremely unlikely and require very low water potentials for their occurrence.

EFFECTS OF CAVITATION IN THE XYLEM

(a) *Gain in water potential*: Suppose cavitation is initiated in a xylem compartment (Fig. 6a). The pressure increases instantaneously to the vapour pressure of water. As a result of the higher water potential in this compartment, a water potential gradient now exists inducing a flow of water from the cavitating cell into the neighbouring cells where the volume and the water potential will increase (Fig. 6b). The extent of the increase in water potential depends on capacities and on the number of cavitating cells.

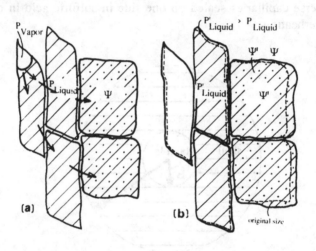

FIG. 6. Increase in the water potential due to cavitation in a xylem vessel. (a) During initiation of cavitation the pressure is increased. The increase in water potential induces a flow of water into neighboring compartments. (b) The water potential and the size of the surrounding tissue is increased. The magnitude of the increase depends on the number of cavitating compartments.

(b) *Increased xylem resistance to water flow*: Cavitation reduces the water conducting cross section of a stem section and, consequently, increases the drop in water potential between roots and plant tops. One would usually expect this drop to be more influential than the previously discussed increase.

EFFECT OF CAVITATION IN LIVING CELLS

(a) *Cavitation is lethal*: (Fig. 7). The entire cell content is emptied into the apoplast. Here again one can visualize several possibilities: i) solutes and water are taken up from the apoplast by adjacent cells.

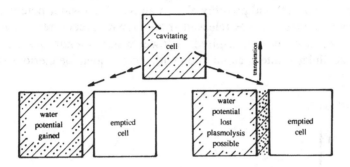

FIG. 7. Theoretical possibilities of the effect of lethal cavitation of a living cell.

(b) *Cavitation is not lethal and the membrane systems remain intact*: (Fig. 8). The pressure in the cavity (and in the entire cell) rises to the level of the vapour pressure, *i.e.* the water potential is increased. A flow of water out of the cell is induced, raising the water potential of the entire tissue. At equilibrium, atmospheric gases will have raised the pressure in the cavity further to nearly that of the free atmosphere and the water potential of the cavitated cell will only have an osmotic component similar to that of a plasmolysed protoplast. The cavitating cell will yield water until it is again in equilibrium with the surrounding cells. Its osmolality will have increased.

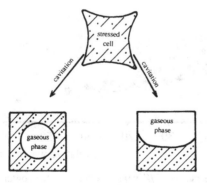

FIG. 8. Theoretical possibilities of the effect of cavitation when the cavitating living cell survives.

This leads to an increase in cellular turgor pressure (because of a lowered osmotic component) and possibly also to an increased water potential; ii) alternately, solutes that are released from cavitating cells remain in the apoplast whilst water is transpired. A lower water potential results from this extracellular solute accumulation with a consequent deleterious effect on turgidity.

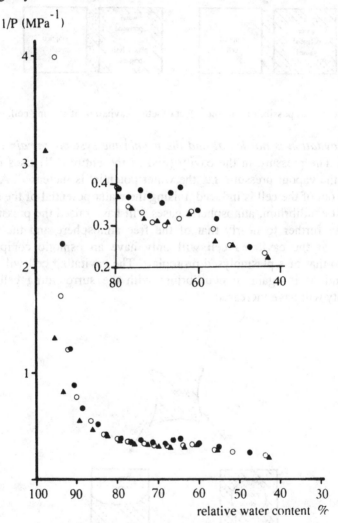

FIG. 9. Pressure-volume curves ($1/P$ versus relative water content) of three different shoots of European beech. The insert shows a section of the same data with an enlarged scale of the ordinate. This section indicates an increase of water potential with each of the three shoots.

DEMONSTRATION OF THE GAIN IN WATER POTENTIAL

(a) *Model calculations.* In model calculations it is possible to show effects of the volume modulus of elasticity for compression, of the stress at which cavitation occurs, of the relative number of cavitating cells and of the osmolality of the cells on the gain of water potential. Unfortunately, a limitation of space does not permit us to present details.

(b) *Evidence from pressure-volume curve.* We obtained pressure-volume curves (plots of $1/P$ versus relative water content) with shoots of the European beech by the method in which the plant shoot is removed from the pressure chamber after each measurement and allowed to lose some water to the free atmosphere before the next measurement (Fig. 9). We often observed that, after reaching the turgor loss point the curves tended to remain horizontal or even to increase for a few measurements before showing the "linear" decrease. We have observed this phenomenon too often to believe in random error and we interpret the data as being evidence of a gain in water potential due to cavitation. This gain in water potential coincides with the appearance of large dry patches of one to several cm in diameter on the leaves. Although we cannot exclude cavitation in the xylem, the appearance of dry spots and the gain in water potential perhaps suggest cavitation of leaf cells. It is clear that the potential gain will not be observed if the shoot is left in the chamber during the entire series of measurements and the pressure is gradually increased.

REFERENCES

Askenasy, E. (1895) Ueber das Saftsteigen. *Botanisches Zentralblatt,* **62,** 237-238.

Böhm, J. (1893) Capillarität und Saftsteigen. *Berichte Deutsche Botanische Gesellschaft,* **11,** 203-212.

Dixon, H.H. & Joly, J. (1895) On the ascent of sap. *Royal Society (London) Philosphical Transactions,* B **186,** 563-576.

Kuroda, K. (1989) Terpenoids causing tracheid-cavitation in *Pinus thunbergii* infected by the pine wood nematode *(Bursaphelenchus xylophilus). Annals of the Phytopathological Society of Japan,* **55,** 170-178.

Lange, O.L. & Lösch, R. (1980) Plant water relations. *Progress in Botany,* **41,** 10-43.

Oertli, J.J. (1989a) The plant cell's resistance to consequences of negative turgor pressure. In *Stuctural and functional responses to environmental stresses* (ed. K.H. Kreeb, H. Richter & T.M. Hinckley), 73-88, SPB Academic Publishing, The Hague, The Netherlands.

40 J.J. OERTLI

Oertli, J.J. (1989b) Die physiologische Bedeutung von Wasserpotentialkomponenten. *Bulletin Bodenkundliche Gesellschaft der Schweiz,* 13, 163-174.

Oertli, J.J., Lips, S.H. & Agami, M. (1990) The strength of sclerophyllous cells to resist collapse due to negative turgor pressure. *Acta Oeologica,* 11, 281-289.

Skaar, C. (1988) *Wood-water relations.* Springer-Verlag, Berlin.

Sperry, J.S. & Tyree, M.T. (1988) Mechanism of water stress-induced xylem embolism. *Plant Physiology,* 88, 581-587.

Tyree, M.T. (1976) Negative turgor pressure in plant cells: fact or fallacy? *Canadian Journal of Botany,* 54, 2738-2746.

Tyree, M.T. & Sperry, J.S. (1989) Vulnerability of xylem to cavitation and embolism. *Annual Review of Plant Physiology, Molecular Biology,* 40, 19-38.

Stomatal control of xylem cavitation

H.G. JONES AND R. A. SUTHERLAND

Horticulture Research International,
Wellesbourne, Warwick CV35 9EF, UK.

SUMMARY

The role of stomatal responses in controlling the occurrence of xylem cavitations is discussed using a simple model of water flow in plants. It is shown that there can be situations where it may, at least in theory, be advantageous to a plant to allow the loss of a proportion of the conducting xylem elements. It is shown that this is a non-trivial conclusion, and further that the achievement of optimal stomatal behaviour may require information on soil water availability and cannot be based on leaf water potential alone.

INTRODUCTION

Although there has been increasing awareness in the past decade of the potential significance of xylem embolisms for plant water relations, there is still an incomplete understanding of the importance of cavitations and consequent xylem embolism as a major factor in plant adaptation to dry environments. In this chapter we attempt to provide a theoretical basis for consideration of the significance of xylem embolisms and the different ways by which different plant species survive and grow in drought conditions. It might be expected that any loss of functional xylem as a result of xylem embolism would be disadvantageous. It is possible, however, that adaptation to dry conditions may involve either mechanisms that minimise the occurrence of cavitations, or mechanisms that minimise their detrimental consequences. In general it seems likely that more 'conservative' behaviours that minimise cavitations (for example by reducing water loss through stomatal closure) will also reduce the chances of high productivity. Circumstances where some controlled cavitation may be advantageous in terms of productivity are identified and appropriate stomatal control functions are outlined.

THEORY AND ASSUMPTIONS

The following treatment largely follows that given by Jones & Sutherland (1991) who used a very simplified hydraulic flow model to investigate stomatal behaviours that might optimise either survival or productivity in relation to two alternative contrasting criteria:

(i) The most conservative behaviour in relation to xylem embolism would be to maintain a maximally efficient conducting system and avoid any loss of conducting vessels. This could be achieved by regulating stomatal closure to prevent the occurrence of water potentials below the threshold needed to cause cavitations. This behaviour should enhance survival, but might tend to reduce photosynthesis and productivity in the short term (though one benefit could be the ability to recover fully when conditions improve).

(ii) An alternative approach could be to maximise production in the short term, even if it involves allowing some embolisms. In what follows we aim to determine whether, and if so under what conditions, maximising productivity might involve some embolism.

Assumptions:

(i) *Water flow model.* Water flow through the soil-plant-atmosphere system is assumed to be described by a simple resistance model with one lumped resistor:

$$\Psi = \Psi_o - E\,R = \Psi_o - E/L \qquad \text{Equation 1}$$

where Ψ is the plant water potential, Ψ_o is the soil water potential, E is the transpiration rate, R is the stem resistance (L, the stem conductance, is the inverse of R). The value of E is given by the product of a constant (c), the stomatal conductance (g_l) and the evaporative demand (D) (*i.e.* $E = c\,D\,g_l$), where D is defined as the water vapour pressure difference between the leaf and the air at the leaf surface. Unfortunately, as well as depending on air humidity, D does depend on environmental factors such as windspeed and radiation, as well as on g_l itself. Nevertheless, for constant environmental conditions D can be approximated by the atmospheric water vapour pressure deficit. The effect of g_l on leaf temperature, and hence on D, can be ignored, at least where D is large.

All conductances and fluxes are expressed per unit ground area.

(ii) *Relationship between conductivity and vessel embolism.* It is necessary to assume a given vulnerability curve relating the number of vessels embolised to the lowest water potential reached. It is recognised that embolisms are most closely related to xylem tension, but this is normally closely related to Ψ in the vessels, which in turn is related to the water potential (Ψ) of the leaves. For convenience the latter is used though it is recognised that the actual Ψ in the conducting vessels will generally be somewhat higher, and the tension correspondingly lower (Tyree & Sperry, 1988). The value of L is assumed to decrease in proportion to the number of vessels embolised, so that L is a function ($L(\Psi)$) of the minimum Ψ reached. Of course the relationship may not be strictly linear because the larger vessels may have a greater probability of embolism, so that because of Poiseuille's Law, the effect on L could initially be disproportionately greater, though this does not substantially affect the following argument. Substituting into Equation 1 we get

$$(\Psi - \Psi_o) = - c D g_1 / L(\Psi) \qquad \text{Equation 2}$$

Given some function $L(\Psi)$ that describes the xylem vulnerability profile with declining Ψ, the problem is to determine the stomatal behaviour $g_1 (\Psi, \Psi_o, cD, ...)$ that achieves each of the optimisation criteria outlined above. In order to illustrate the general principles it is appropriate to approximate observed data sets (Fig. 1) by the following linear function as shown in Fig. 1b:

for $\qquad \Psi_t < \Psi \qquad L(\Psi) = L_{max}$,

for $\qquad \Psi_t > \Psi > -X \qquad L(\Psi) = L_{max} (\Psi + X)/(X + \Psi_t)$,

for $\qquad -X > \Psi \qquad L(\Psi) = 0$

$$\text{Equation} \quad 3$$

where L_{max} is the maximum hydraulic conductance, Ψ_t is the threshold potential at which L starts to decline and $X = -\Psi$ at which L falls to zero.

More sophisticated analyses will need to take account of actual functions relating L and Ψ. For illustrative purposes an L_{max} of 0.005 mol MPa^{-1}m^{-2}s^{-1} (approximately 9×10^{-8} mMPa^{-1}s^{-1}) has been used for all calculations. (This would give a soil-to-leaf water potential difference of 1 MPa for a typical high crop evaporation rate (Jones, 1983) of 0.005 mol m^{-2} s^{-1} or 0.35 mm h^{-1}.)

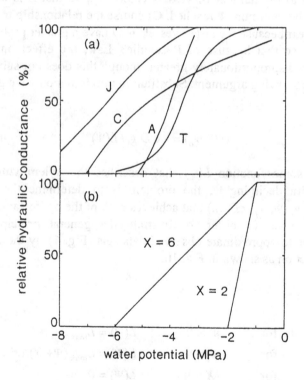

FIG. 1. (a) Typical xylem vulnerability curves for *Cassipourea elliptica* (Sw.) Poir. (C), *Juniperus virginiana* Marsh. (J), *Acer saccharum* L. (A) and *Thuja occidentalis* L. (T) from Sperry, Tyree & Donnelly (1988), Sperry & Tyree (1990), and Tyree & Dixon (1986); (b) two model vulnerability curves (from Jones & Sutherland, 1991).

(iii) *The effect of low water potential is primarily on air entry into vessels.* This assumption implies that physiological processes are unaffected by leaf water potential over the full range above that causing complete vessel collapse (Ψ_{min}). Evidence for this assertion includes: physiological processes tend not to be related to Ψ *per se,* but to turgor (Jones, 1973), while turgor can be maintained at increasingly low values of Ψ by osmotic adjustment (Turner & Jones, 1980); the rate of photosynthesis at low Ψ is determined primarily by stomatal closurerather than by direct damage to the photosynthetic biochemistry by the low Ψ (Farquhar & Sharkey, 1982); there is increasing evidence that even stomatal closure is not regulated by leaf Ψ, but by soil/root water status (see Davies & Jeffcoat, 1990; Jones, 1990).

(iv) *Embolisms are essentially irreversible.* The plant is assumed to be non-growing, so embolisms are unlikely to be reversed within a growing season (Milburn, 1979; Tyree & Sperry, 1989).

(v) *Productivity is assumed to be proportional to stomatal conductance.* Stomatal conductance (g_l) can vary between zero and a maximum value ($g_{l\ max}$), with assimilation being approximately proportional to g_l (*e.g.* Wong, Cowan & Farquhar, 1985) even as water potential falls. Productivity would therefore be approximately proportional to g_l, though there are a number of factors that might cause such a relationship to be only approximate.

RESULTS AND DISCUSSION

The model was used to investigate the stomatal behaviours required to satisfy the two alternative conditions outlined above.

(i) Consider first the behaviour that would be required to avoid any embolism, yet maximises stomatal aperture and hence productivity. In this case Ψ can only be in the range 0 to Ψ_t where Ψ_t is the value above which cavitations do not occur. That is, Ψ must be equal or greater than Ψ_t, so, substituting into Equation 2, remembering that $L(\Psi) = L_{max}$ to satisfy the criteria set, we get the condition that:

$$g_l <= -\,(\Psi_t - \Psi_o)\,L_{max}\,/cD \qquad\qquad \text{Equation 4}$$

This limiting behaviour for g_l is illustrated by the solid lines in Fig. 2a, which shows that as demand (*cD*) increases from zero, the stomata stay maximally open until a threshold value of demand is reached. This threshold demand for initiation of stomatal closure is smaller as the soil water potential (Ψ_o) decreases. Below this threshold the stomata close hyperbolically with increasing demand. In each case stomata start to close as soon as Ψ falls to Ψ_t (in this example -1.0 MPa).

(ii) Alternatively we can consider the requirement to maximise g_l, even if this may involve some loss of functional xylem. In this case we need to find the value of Ψ ($= \Psi_{to}$) that maximises g_l in Equation 2 (i.e. find the value of Ψ_{to} that maximises $-(\Psi_{to} - \Psi_o)L(\Psi)/cD$).

Differentiating this expression with respect to Ψ_{to} and setting the result equal to zero will give the minimum value of Ψ at which $g_{l\,max}$ can be maintained. Substituting for L_{max} from Equation 3 and differentiating gives

$$\Psi_{to} = (\Psi_o - X)/2 \qquad\qquad \text{Equation 5}$$

Because no damage occurs for $\Psi > \Psi_t$ it follows that the critical value of Ψ_{to} is min $\{\Psi_t, (\Psi_o - X)/2\}$. From inspection it is apparent that where the slope of the vulnerability profile is not too steep, g_l is maximised by allowing Ψ to fall below the critical value at which cavitations occur with the consequent occurrence of some cavitation. With the parameterisation chosen, for example, and when $\Psi_o = 0$, occasions where g_l is maximised by allowing some vessel loss occur for any $X > 2$. The optimal behaviour of g_l is illustrated for $X = 6$ in Fig. 2 for a range of values of Ψ_o. The corresponding changes in L are presented in Fig. 2b and values for Ψ are presented in Fig. 2c (calculated as outlined in Jones & Sutherland, 1991). Variation of Ψ_{to} with Ψ_o is also given in Jones & Sutherland (1991), where it is shown that Ψ_{to} decreases with Ψ_o and with increasing X. In the region where $\Psi < \Psi_{to}$ there must be some stomatal closure in order to avoid catastrophic xylem failure. It is also necessary to note that the trajectory of Ψ is not reversible, in that after some vessels have been lost, Ψ will not recover along the same line as was followed as stress increased (Fig. 2c).

It must be recognised that throughout the above treatment it has been necessary to make a number of simplifications. For example, in addition to those mentioned above it is worth noting that the differentiation used to obtain Equation 5 assumed that D is constant, though in practice it may be slightly dependent on g_l.

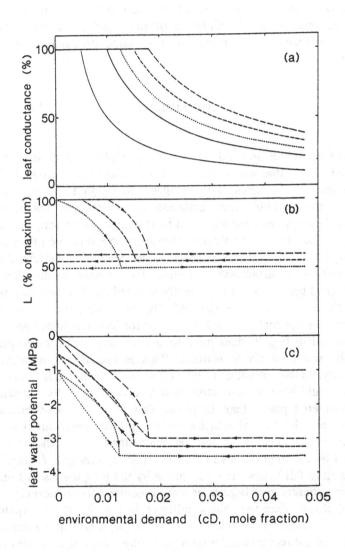

FIG. 2. Dependence on environmental demand of (a) leaf conductance, (b) L, and (c) leaf water potential. Solid lines represent the safety strategy (Case i) with $\Psi_o = 0$ MPa. Dashed lines represent Case ii (with $X = 6$, $\Psi_t = -1.0$) with the dash length decreasing from $\Psi_o = 0$ through -0.5 to -1.0 MPa (after Jones & Sutherland, 1991).

The error will normally be small, especially for a large area of crop poorly coupled to the environment (McNaughton & Jarvis, 1983). The objective of the present work is simply to investigate the general consequences of different types of stomatal behaviour in relation to embolism, more detailed validation of the conclusions for particular situations will require more precise analysis using more realistic response functions.

CONCLUSIONS

The most important and indeed surprising implication of the analysis presented here is that where the criterion is maximising productivity, it can in some circumstances be advantageous to a plant to have a stomatal behaviour that permits some cavitation and consequent loss of xylem function. This applies even where, at least within one season, cavitations are irreversible. It is of particular interest to note that even though this response leads to irreversible 'damage' to the conducting system, there is no consequential disadvantage (under the present assumption that physiological processes are not directly affected by Ψ) when the stress is caused by high evaporative demand. On the other hand, it should be noted that the optimal stomatal behaviour for low soil water potentials, as illustrated in Fig. 2, does have long-term detrimental consequences when the water supply is restored. This is because some additional vessels beyond the optimum at high soil water are lost and it follows that stomata would have to start closing at a lower evaporative demand in these rewatered plants than in plants that have not been previously stressed. In order to maintain long-term optimal behaviour when soil water potential falls, Ψ should be maintained above the minimum indicated by the long dashes in Fig. 2c, (correspondingly, L should not be allowed to fall below the value given by the long dashes in Fig. 2b). This more conservative behaviour would require correspondingly earlier stomatal closure than the values indicated in Fig. 2a, at the expense of some loss in short term productivity. Of course, a critical assumption is that Ψ is not a direct constraint on normal physiological processes (other than through its effect on cavitation).

It is important to remember that other constraints may be important in practice, such as the possible need to conserve water for later use with the consequent possibility of an optimal stomatal conductance below that derived by this analysis (Jones, 1981).

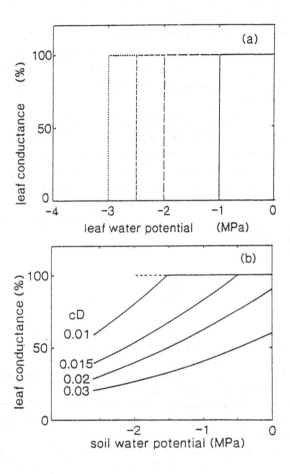

FIG. 3. (a) Relationship between leaf conductance and leaf water potential for the standard model ($X = 6$) with symbols as for Fig. 2. (b) Relationship between leaf conductance and soil water potential as demand changes (for Case ii).

The relationships between leaf conductance and leaf water potential for the two main stomatal behaviours described above are illustrated in Fig. 3a. It is apparent from this that both the safety strategy and the maximal g_l strategy require abrupt stomatal closure as Ψ falls to a particular value, which for the maximal g_l strategy is itself a function of Ψ_o. This abrupt stomatal closure has important implications for the control of g_l. In particular there is no signal relating to Ψ that can give information on the need for intermediate values of g_l. This implies that any optimal control of g_l cannot be achieved in terms of Ψ.

Similarly, Fig. 2 shows that optimal control cannot be achieved in terms of the evaporative demand (cD) alone. If one plots g_l as a function of Ψ_o (Fig. 3b), it is apparent that it is possible to achieve optimal control through action of both Ψ_o and cD. Alternatively it would be possible to achieve the same objective through a combination of Ψ_o and Ψ.

The main conclusions from this latter analysis of the control possibilities indicates that a knowledge of Ψ_o is essential to obtain optimal control, and that a knowledge of leaf Ψ is **not** sufficient. This provides further supporting evidence, though only circumstantial, that a plant should be able to sense and respond to Ψ_{soil} if it is to behave optimally (Jones, 1981), and confirms experimental evidence that stomata may respond to Ψ_{soil} rather than to Ψ_{leaf} in many situations (*e.g.* Gollan, Passiuora & Munns, 1985).

REFERENCES

Davies, W.J. & Jeffcoat, B. (eds.) (1990) *Importance of Root to Shoot Communication in the Responses to Environmental Stress.* British Plant Growth Regulator Group, Long Ashton, Bristol.

Farquhar, G.D. & Sharkey, T.D. (1982) Stomatal conductance and photosynthesis. *Annual Review of Plant Physiology,* 33, 317-345.

Gollan, T., Passioura, J. & Munns, R. (1985) Soil water status affects the stomatal conductance of fully turgid wheat and sunflower leaves. *Australian Journal of Plant Physiology,* 13, 459-464.

Jones, H.G. (1973) Photosynthesis by thin leaf slices in solution. II. Osmotic stress and its effects on photosynthesis. *Australian Journal of Biological Sciences,* 26, 25-33.

Jones, H.G. (1981) The use of stochastic modelling to study the influence of stomatal behaviour on yield-climate relationships. In *Mathematics and Plant Physiology* (ed. D.A. Rose & D.A.Charles-Edwards), pp. 231-244. Academic Press, London.

Jones, H.G. (1983) *Plants and Microclimate.* Cambridge University Press, Cambridge.

Jones, H.G. (1990) Physiological aspects of the control of water status in horticultural crops. *HortScience,* 25, 19-26.

Jones, H.G. & Sutherland, R.A. (1991) Stomatal control of xylem embolism. *Plant, Cell & Environment,* 14, 607-612.

McNaughton, K.G. & Jarvis, P.G. (1983) Predicting effects of vegetation changes on transpiration and evaporation. In *Water Deficits and Plant Growth, vol VII* (ed. T.T.Kozlowski), pp. 1-47. Academic Press, New York and London.

Milburn, J.A. (1979) *Water Flow in Plants.* Longman, London.

Sperry, J.S. & Tyree, M.T. (1990) Water-stress-induced xylem embolism in three species of conifers. *Plant, Cell & Environment,* 13, 427-436.

Sperry, J.S., Tyree, M.T. & Donnelly, J.R. (1988) Vulnerability of xylem to embolism in mangrove vs. inland species of Rhizophoraceae. *Physiologia Plantarum,* 74, 276-283.

Turner, N.C. & Jones, M.M. (1980) Turgor maintenance by osmotic adjustment: a review and evaluation. In *Adaptation of Plants to Water and High Temperature Stress* (eds. N.C.Turner & P.J.Kramer), pp. 87-103. Wiley, New York.

Tyree, M.T. & Dixon, M.A. (1986) Water stress induced cavitation and embolism in some woody plants. *Physiologia Plantarum,* 66, 397-405.

Tyree, M.T. & Sperry, J.S. (1988) Do woody plants operate near the point of catastrophic xylem dysfunction caused by dynamic water stress? *Plant Physiology,* 88, 574-580.

Tyree, M.T. & Sperry, J.S. (1989) Vulnerability of xylem to cavitation and embolism. *Annual Review of Plant Physiology and Molecular Biology,* 40, 19-38.

Wong, S.C., Cowan, I.R. & Farquhar, G.D. (1985) Leaf conductance in relation to rate of CO_2 assimilation. III. Influence of water stress and photoinhibition. *Plant Physiology,* 78, 830-834.

Refilling of embolized xylem

J. GRACE

Institute of Ecology & Resource Management,
The University of Edinburgh, Darwin Building,
Mayfield Road, Edinburgh EH9 3JU, UK.

SUMMARY

The water content of xylem has often been shown to undergo seasonal and even diurnal fluctuations, apparently in response to changes in transpirational flux. When the water content declines, it may be the result of cavitation in water in tracheids or simply indicative of water menisci receding into the tapered ends of already-cavitated tracheids. Similarly, an increase in water content may be caused by movement of menisci and/or complete refilling of tracheids. Recent experiments suggest that parenchyma has no role in the refilling process. Rather, refilling is brought about by the dissolving of trapped gas in the tracheids as the menisci adjust their position in response to the water potential.

INTRODUCTION

It has long been known that the water content of the woody tissues of trees undergo seasonal fluctuations, but the fact was overlooked in most discussion of the theory of water transport and little attention was given to the underlying processes. Some of the most convincing early data illustrating the phenomenon were provided by Gibbs, working on Canadian forest trees in the 1930s (Gibbs, 1958). He showed that the wood of young trees of *Betula populifera* underwent an annual cycle of water content, varying from 100 per cent of dry weight in the early spring to as little as 60 per cent in the late summer (Fig. 1). Fluctuations were most extreme, and earlier in the season, in the upper part of the stem; and the rise in the spring was immediately after the thawing of the soil. The magnitude of the fluctuations seem particularly large when it is realised that a large part of the weight of woody tissue is the cell wall: if the data were to be re-expressed as a water fraction and a gas fraction, it would be clear that a reduction to 60 per cent of the dry weight represents a very large loss of water content indeed.

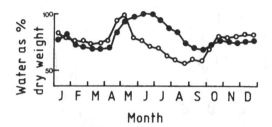

FIG. 1. Seasonal changes in water content of *Betula populifera* according to Gibbs (1958). Tops and middles of the stems are indicated by dots and circles respectively.

Difficulty in reconciling such data with the cohesion theory of water transport arises because if the transport system becomes gas-filled when the water is under tension, then continuity of the transpiration stream is lost. In the least damaging case, this might merely reduce the hydraulic conductance of the xylem, but at worst embolism might occur throughout the xylem. However, quite early on, it was shown that many species continue to transpire and transport water, even when saw-cuts are made on opposite sides of the trunk to reduce the conducting area. Radio-isotopes and dyes used as tracers in such experiments, enable us to see that water is transported laterally, around the cuts; and the conclusion has often been reached that plants have 'spare capacity' to transport water (Kozlowski, 1965), or that cavitation is often of adaptive value, enabling the tree to draw upon stores of water in the xylem (Zimmermann, 1983). A contrasting view, is that the xylem is a 'vulnerable pipeline', and that any reduction in the conducting area of the sapwood will result in an decreased hydraulic conductance, and thus a decline in leaf water potential (Milburn, 1979).

Fluctuations in xylem water content over periods of weeks have been shown by Chalk & Bigg (1958), Roberts (1976), Waring & Running (1978) and Waring, Whitehead & Jarvis (1979). In mature Douglas Fir, the water content of the sapwood was seen to fluctuate with the calculated potential transpiration rate (Fig. 2). It appears from these data that emptying and refilling of xylem occurs over periods of a few weeks. Emptying occurs whenever transpiration is high, and refilling occurs when the transpiration is low. Even more rapid changes occur in the data on apple trees of Brough, Jones & Grace (1986): the water content fluctuated in direct response to the water potential over the course of a single day, falling in the morning, and increasing in the afternoon.

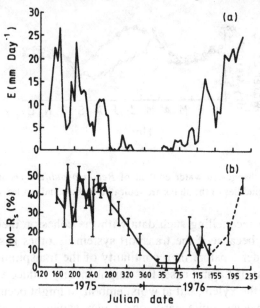

FIG. 2. Relationship between sapwood water deficit (100-R_s) and calculated potential evapotranspiration (E) for Douglas Fir. Redrawn from Waring & Running 1978).

POSSIBLE INTERPRETATIONS OF THESE DATA

1. Declining water content

When the water content of wood declines, cavitation may or may not occur. Many woody tissues contain substantial amounts of living parenchyma cells, and possibly 'capillary water' trapped between adjacent tracheids and inside fibres (Zimmermann, 1983). Many tissues are to some extent elastic, and their water-containing volume will therefore change with pressure. Angiosperm trees have woody tissues which are up to 10% by volume parenchyma. Thus, there are several reasons why woody stems might be expected to fluctuate in their water content as a complex function of water potential, without any cavitation occuring. However, the maximum fraction of xylem water that can be accounted for by shrinking and swelling, and by changes in capillary water is relatively small, perhaps only 5% for most gymnosperms, and somewhat higher for angiosperms. Moreover, acoustic sensing of woody tissues shows that cavitation often does occur when the water potential falls below -1.5 MPa (Dixon, Grace & Tyree, 1984; Peña & Grace, 1986; Tyree & Sperry, 1989), but this may be a feature of the early summer. Later, all tracheids or vessels which have cavitated may remain cavitated, the remainder being less vulnerable to cavitation.

In reviewing the early literature it should also be remembered that good measurements of sap water content are hard to obtain. When cores are taken from actively transpiring trees, the tension in the transport system means that embolism may occur when the samples are cut, a point acknowledged by Gibbs (1958). Specially designed cutters to make two simultaneous cuts may help to prevent this from occuring. Perhaps the best method of assessing the water content of tree trunks is the gamma probe (Edwards & Jarvis, 1981). This device measures the attenuation of a collimated beam of radiation which traverses the stem. Recently, it has been possible to scan trunks of mature trees in several directions, and using computer tomography to map the distribution of water inside the section (Habermehl, 1982; Zainalabidin & Crowther, personal communication). This technique, applied to park trees in Glasgow, shows beyond any reasonable doubt that very substantial regions of the transport system are cavitated in the summer.

What actually happens inside the tracheid and vessel following cavitation? Immediately after the event, it is supposed that the tracheid contains water vapour in equilibrium with the liquid water which exists at the ends of the tracheids and attached to the hydrophilic walls. Gases from the outside air, and carbon dioxide from respiration of ray parenchyma cells will diffuse into the void, assisted by the fact that the surrounding water is not stationary but moving quite rapidly upwards. Thus, the tracheid comes to be gas filled, with a water potential equal to that of the surrounding tissues, the gas pressure being in equilibrium with the outside air. There are very few measurements of the gas composition, but what there are suggest it is CO_2-enriched air (Sperry et al., 1987).

2. Increasing water content

The refilling process requires that the gas be redissolved in the water. Of the major constituent gases, all are soluble in water but nitrogen is the least soluble (0.015 m^3 gas per m^3 of water, Kaye & Labey, 1973). Gas bubbles in slices of air dried wood can be observed to dissolve in free water on the stage of a microscope. But when the water is under tension, even if dissolution does occur, the resulting low pressure would not be enough to bring the menisci together. In some species, the refilling occurs when the xylem tension is abolished by root pressure. Sperry *et al.*(1987) measured the root pressures developed in grapevines, using manometers attached to the roots. Refilling was assessed by measuring the hydraulic conductivity. These authors found root pressures during spring refilling to be 10-100 kPa, and concluded that under these pressures, the gas would have simply dissolved in the water. Root pressures can be much higher than this: Hales in 1727 measured substantial root pressures in grape, and pressures of 0.6 MPa have been recorded in tomato. However, these pressures are in detopped plants, and do not occur when the plant is actively transpiring. In the summer months, xylem of north temperate trees remains under tension, even at pre-dawn, so the refilling noted in the Douglas Fir data (Fig. 2) cannot be attributed to root pressure.

According to one theory, let us call it the active-osmosis theory, the ray parenchyma has a role in maintaining the water balance of the xylem, and may be responsible for refilling. Anatomical studies show that in conifers each tracheid is in contact with the medullary ray tissue (Carlquist, 1975), so that entry of water into a tracheid may be influenced by metabolic activity of the associated parenchyma cell (Wodzicki & Brown 1970; Braun, 1983). It is unclear how parenchyma could perform this role, as living cells do not generally pump water directly, but move water by pumping cations and thus creating a gradient of water potential. The mechanism envisaged requires the water potential of the small volumes of water in the tracheids to be lowered by cations (Fig. 3).

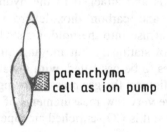
parenchyma
cell as ion pump

FIG. 3. A postulated refilling mechanism: the parenchyma cell pumps cations into the residual water, causing a decline in the solute potential of the sap.

Then, water from surrounding cells would flow into the cell, putting the gas under a positive pressure, enabling it to dissolve. Cell walls of tracheids are permeable to water according to the measurements of Petty & Palin (1983).

These authors concluded that the filling of a wood cell might be possible in as little as 5 minutes under a pressure difference of 0.1 MPa if other flow pathways (pits) were blocked.

In this hypothesis, the ray cell must be living and capable of secreting appropriate ions, and possibly transporting further ions from other regions of each individual ray. Recently, in our glasshouse, the hypothesis was tested using *Pinus sylvestris*. Young trees, 2-m tall, were instrumented with acoustic sensors and their stem water content was recorded using a gamma probe (Borghetti *et al.*,1991). Their roots were removed from soil and placed in sand whilst the leaf water potential and stem water content was measured. Over a six-day period, the water potential fell to -2.5 MPa, the water content of the stem fell and acoustic emissions were detected (Fig. 4). After that, the bases of stems were recut under water, and the stems were stood in (a) water or (b) water containing 10 mol m^{-3} of the respiratory poison, sodium azide. This strength of azide had previously been shown to kill xylem. Despite the poison, rehydration occured equally in both the azide and non-azide treatments. In the rehydration period, the water potential of the leaves was -0.5 MPa. Unfortunately, the potential of the xylem was not measured, but it is presumed to have been midway between that of the water supplied and the foliage.

Other experiments were conducted to test whether azide affected the rate of hydration of small segments of stem immersed in water. Another series of experiments on potted plants sought to kill the parenchyma by enclosing the stem in a special collar, through which hot water was circulated. Both sorts of experiment lead us to the conclusion that refilling did not depend on the parenchyma cells being alive (Borghetti *et al.*,1991). We did however find that xylem sap from water stressed plants contained more osmotically active substances than those from well watered plants (osmotic potential was -0.11 and 0.07 MPa respectively). The consequent decline in water potential of the bulk xylem sap might have masked a considerable local decline in those regions where the cells are cavitated. So we cannot exclude the possibility than the action of the parenchyma cells *before they were killed* could have contributed appreciably to their refilling rate.

A second theory, which does not require any living process, is the capillary action theory. According to this, the capillary effect of traces of water left in the ends of a tracheid reduces the water potential below that of adjacent cells. The water potential Ψ can be estimated from the radius of the meniscus, using the well-known formula (Skaar, 1972):

$$\Psi = (2 \, \sigma) \, / \, r \qquad\qquad \text{Equation 1}$$

where σ is the surface tension of water (7.3×10^{-2} Pa m at 293 K), and r is the radius of the gas-liquid interface.

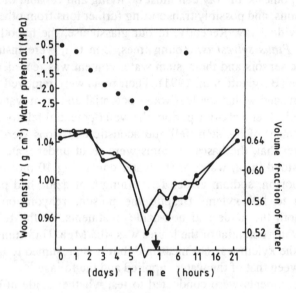

FIG. 4. Variation in xylem water potential and stem water content of young *Pinus sylvestris* trees during the drought phase and after resupply of water. Note the change of scale during the rehydration phase (water supply given where the arrow is shown). In the case of the solid circles, the water used to rehydrate the tissues was poisoned with sodium azide. The arrow denotes when water was resupplied. Redrawn from Borghetti *et al.*(1991).

For a partially-filled tracheid with a water/air meniscus of radius of 15 µm, a water potential of -9.7 kPa can be expected, but when the water occupies the pointed ends of the tracheids the radius is much smaller and the potential will be much less (i.e. more negative). On the basis of this, it is possible to calculate a relationship between water content and water potential, based on the geometry of the cell; in fact, a pressure volume curve for a single tracheid. Tracheids are usually quite long, and the pointed ends are only a small part of their volume. We would therefore expect that the pressure volume curve would show a very small change in potential over a wide range of dehydration, becoming very negative when the last traces of water were being removed. Pressure volume curves of

conifer wood have been determined by Waring *et al.* (1979) and Edwards
& Jarvis (1982). These curves are determined by allowing pieces of wood
to dry to equilibrium over a salt solution of known osmotic potential.
They do indeed show a wide range of water content over which the
potential declines only a little (Fig. 5). The water content, as relative
water content, is seen to fall from 75 to 35% with a decline in water
potential of only -0.5 MPa. Thereafter, the potential falls sharply with
very little reduction in water content (Fig. 5). The water potentials in the
physiological range (100-35%) do however, fall more than one would
have expected from the radius of the tracheids, perhaps because of the
addition of an osmotic component and a surface (wall) component.

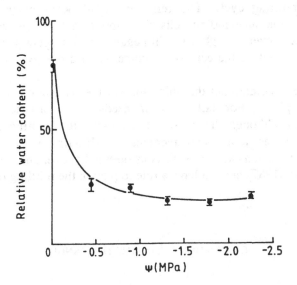

FIG. 5. Relative water content of *Pinus sylvestris* sapwood equilibrated over salt
solutions at a range of water potentials. Redrawn from Edwards and Jarvis (1982).

Predawn water potentials in the summer months are not well
characterised, as interest has usually centred on the daytime values, and
the measurement has usually be made on foliage rather than in the stem.
These values are typically -0.1 to -0.5 MPa. The fact that refilling can
occur when water is applied under tension can be demonstrated by the
experiment described by Borghetti *et al.*(1991). These authors connected
fine plastic tubes to partially dried stem segments, which had undergone

cavitation. The tubes were allowed to hang in flasks of water resting on a balance. The uptake or efflux of water from the segment could then be measured. Experiments of this type in our laboratory have shown that water can be taken up when the flasks are lowered to 3 m below the level of the segment, corresponding to a water potential of 30 kPa. Data are lacking on the exact water potentials occuring in the trunks of trees during summer refilling, and most measurement systems (thermocouple psychrometers and pressure chambers) are not sufficiently accurate in this range. Moreover, there have been few measurements of xylem osmotic potential in trees, and it is unclear how far the xylem potential might be reduced by leakage from the parenchyma into the tracheid.

On the basis of the above discussion, it is possible to postulate a cavitation/refilling cycle. The refilling would occur when the water potential in the surrounding cells rises above that of the water in the cavitated cell (even though it is still negative with respect to free water). Then, the gas inside the cell would come under a positive pressure, and dissolve.

It has been pointed out that this capillary mechanism could only work for small plants. For tall trees, it needs to be assisted by other mechanisms, although the possibility of rainwater entering the system from above has long been recognised (Milburn, 1979). Candidate mechanisms, such as active-osmosis as outlined above, and root pressure (Sperry et al.,1987) may all have a role to play in the refilling of xylem in tall trees.

CAVITATION AND CLIMATE CHANGE

It is surprising that we still know relatively little about the ecological significance of cavitation and refilling. In the context of climate change, woody vegetation is likely to experience an increase in the frequence of drought, and overall higher temperatures. Both of these are may increase cavitation and reduce refilling. In drought, the water in the xylem is liable to experience greater tensions, and so exceed the threshold tension more often. But the threshold tension may be lower, as the increase in respiration of the parenchyma and bark tissues may increase the gas content of the xylem water. Whilst there may be a 'normal' extent of cavitation that trees experience and can tolerate, a change in the climate so that warm dry episodes are more frequent and more extreme, may push trees beyond their adaptive range, leading to a new tree decline.

REFERENCES

Borghetti, M., Edwards, W.R.N., Grace, J., Jarvis, P.G. & Raschi, A. (1991) The refilling of embolised xylem in *Pinus sylvestris* L. *Plant, Cell & Environment*, 14, 357-369.
Braun, H.J. (1983) Zur dynamic des wassertransportes in Baumen. *Bericht der Deutschen botanischen Gesellschaft*, 96, 29-47.
Brough, D.W., Jones, H.G. & Grace, J. (1986) Diurnal changes in water content of the stems of apple trees, as influenced by irrigation. *Plant, Cell & Environment*, 9, 1-7.
Carlquist, S. (1975) *Ecological Strategies of Xylem Evolution.* University of California Press, Berkeley.
Chalk, L. & Bigg, J.M. (1958) The distribution of moisture in the living stem of Sitka spruce and Douglas Fir. *Forestry*, 29, 5-21.
Dixon, M.A., Grace, J. & Tyree, M.T. (1984) Concurrent measurements of stem density, leaf water potential and cavitation on a shoot of *Thuja occidentalis* L. *Plant, Cell & Environment*, 7, 615-618.
Edwards, W.R.N. & Jarvis, P.G. (1981) A method for measuring radial differences in water content of intact tree stems by attenuation of gamma radiation. *Plant, Cell & Environment*, 6, 225-260.
Edwards, W.R.N. & Jarvis, P.G. (1982) Relations between water content, potential and permeability in stems of conifers. *Plant, Cell & Environment*, 5, 271-277.
Gibbs, R.D. (1958) Patterns in the seasonal water content of trees. In *The Physiology of Forest Trees* (ed. by K.V. Thimann), pp. 43-69, Ronald Press, New York.
Habermehl, A. (1982) A new non-destructive method for determining internal wood condition and decay in living trees. 2. Results and further developments. *Arboricultural Journal*, 6, 121-130.
Kaye, G.W.C. & Labey, T.H. (1973) *Tables of Physical and Chemical Constants*, 14th edition. Longman, London.
Kozlowski, T.T. (1965) *Water Metabolism in Plants.* Harper & Row, New York.
Milburn, J.A. (1979) *Water Flow in Plants.* Longman, London.
Peña, J. & Grace, J. (1986) Water relations and ultrasound emissions before, during and after a period of water stress. *New Phytologist*, 103, 515-524.
Petty, A. & Palin, M.A. (1983) Permeability to water of the fibre cell wall material of two hardwoods. *Journal of Experimental Botany*, 34, 688-693.
Roberts, J. (1976) An examination of the quantity of water stored in mature *Pinus sylvestris* L. trees. *Journal of Experimental Botany*, 27, 473-479.
Skaar, C. (1972) *Water in Wood.* Syracuse University Press, New York.
Sperry, J.S., Holbrook, N.M., Zimmermann, M.H. & Tyree, M.T. (1987) Spring filling of xylem vessels in wild grapevine. *Plant Physiology*, 83, 414-417.
Tyree, M.T. & Sperry, J.S. (1989) Vulnerability of xylem to cavitation and embolism. *Annual Review of Plant Physiology and Molecular Biology*, 40, 19-38.
Waring, R.H. & Running, S.W. (1978) Sapwood water storage: its contribution to transpiration and effect upon water conductance through the stems of old-growth Douglas Fir. *Plant, Cell & Environment*, 1, 131-140.

Waring, R.H., Whitehead, D. & Jarvis, P.G. (1979) The contribution of stored water to transpiration in Scots pine. *Plant, Cell & Environment*, **2**, 309-317.

Wodzicki, T.J. & Brown, C.L. (1970) Role of xylem parenchyma in maintaining the water balance of trees. *Acta Societatis Botanicorum Poloniae*, **39**, 617-621.

Zimmermann, M.H. (1983) *Xylem Structure and the Ascent of Sap.* Springer, New York.

Interpretation of the dynamics of plant water potential

M. A. DIXON AND R.W. JOHNSON

*Dept. of Horticultural Science, University of Guelph,
Guelph, Ontario, Canada, N1G 2W1.*

SUMMARY

Plant stem water potential was monitored continuously and non-destructively on both woody and herbaceous species under partially controlled greenhouse environment and/or field conditions. The *in situ* stem psychrometer was automated under computer control which resulted in 30 minute time resolution on measurements of stem water potential. These data provided greater time resolution of plant water potential than has been previously attained. Concurrent measurements of ambient vapour pressure deficit and solar radiation were also made. Interactions among physiological and environmental variables were interpreted with respect to specific and varietal differences and the extent to which mechanisms of response to water stress could be demonstrated.

Field measurements of stem water potential on tomato (*Lycopersicon esculentum* L.Var.) implied distinctive varietal differences in the pattern of stomatal response to routine diurnal water stress. Greenhouse experiments on potted tree saplings (*e.g. Fraxinus pennsylvanicum, Gleditsia enermis, Betula verucosa*) exhibited direct correlations between stem water potential and incoming radiation. Clear differences were exhibited between species with respect to their water status under the same environmental conditions. The relationship between stem water potential and stomatal conductance in red ash indicated that partial stomatal closure was induced when the water stress dropped to about -1.0 MPa. This was followed by some rehydration, an increase in stomatal conductance and a second dehydration phase. The kinetics of water potential changes during progressive dehydration and recovery upon rewatering were observed. The automated *in situ* stem psychrometer emerged as a reliable and insightful tool for evaluating plant-environment interaction.

INTRODUCTION

Measuring the responses of plants to environmental stimuli has been the focus of a very large body of research for as long as scientists have contemplated the nature of plant systems. It is certainly true that these responses are often complex and inter-related in such a manner as to confound reliable interpretation. It is equally true that environmental variables influencing plant systems are complex and interactive to the point of further confounding the evaluation of plant-environment interaction. This paper defines an approach to the interpretation of plant-environment interaction based on an analysis of plant water relations and related physiological and environmental variables. The approach comprised a necessarily extensive technical development centred mainly on the *in situ* stem psychrometer (Dixon & Tyree, 1984) for measuring plant water potential.

The thermocouple psychrometer (Spanner, 1951) has enjoyed a wide audience. The attraction of continuous, non-destructive measurements of plant water potential has led to the exploitation of this technique for *in situ* applications (Neumann & Thurtle, 1972; Michel, 1977; McBurney & Costigan, 1982; Dixon & Tyree, 1984). In particular, the temperature corrected, *in situ* stem psychrometer has been further developed and extensively and successfully applied to a variety of plant systems (Dixon, Grace & Tyree, 1984; Dixon, 1987; Coffey, 1989; Darlington, 1990). The operation of this instrument has been automated to allow for greater time resolution of water potential measurements.

Automation of reliable and accurate measurements of plant water potential permits a degree of scrutiny of plant water relations hitherto not available to this field of research. This now affords us the opportunity to examine the kinetics of water status changes in plant tissues in response to various environmental stimuli.

The broader aim of this research is to exploit the plant as an environmental sensor. The long-term objectives are based on the notion that plants routinely integrate the complex interactions of environmental variables and measuring their response in terms of water relations parameters (specifically water potential) may provide an accurate and reliable means to evaluate mechanisms of plant-environment interaction. Within the scope of this study, the reliability of the automated psychrometer for continuous water potential measurement was determined. Also, the sensitivity and interpretation of water potential fluctuations in response to environmental variables was assessed.

MATERIALS AND METHODS

Three species of deciduous trees were potted as whips (1.5 - 2 m long) in 10 litre pots using a standard potting mix (ProMix BX). The species included silver birch (*Betula verrucosa*), thornless honey locust (*Gleditsia inermis*) and red ash (*Fraxinus pennsylvanicum*). Plants were established in the pots in March and grown in a greenhouse where experiments commenced in mid May. Control and monitoring of the environment was achieved with a DGT Model 1220 Climate Control System (Dansk Gartneri Teknik, Denmark). The extent to which environment control (particularly temperature) was successful depended on the time of year, the cooler months exhibiting the best control. Experiments carried out in the summer months could be subjected to variable ambient temperatures in excess of 35 °C.

Two cultivars of processing tomato (*Lycopersicon esculentum*), LA1563 and HY9464 were established in field sites at the Cambridge Research Station, Ontario, Canada. Environmental data for plants grown in the field were also monitored.

Treatments varied from maintenance of well-watered conditions to progressive drought and recovery. Seasonal variation provided a range of vapour pressure deficits and solar radiation levels throughout which changes in water status were monitored continuously.

Temperature, humidity and solar radiation were monitored continuously using electronic sensors attached to a computer. Soil moisture, in the cases of the potted trees, was controlled by continuous monitoring of pot weight, using automated electronic balances, and supplying measured quantities of reverse osmosis treated water and liquid fertilizer as needed.

In situ stem psychrometers (P.W.S. Instruments, Inc., Guelph, Ontario) were used to measure stem water potential and were automated using a datalogger (Model CR7X, Campbell Scientific, Logan, Utah). Normally, as many as twenty stem psychrometers were operated simultaneously on replicate samples under the same environmental conditions. Sample preparation has been described by Dixon (1987). Psychrometer installations were insulated with a 25 mm layer of styrofoam (SM) covered with reflective foil.

In some cases, concurrent measurements of leaf stomatal conductance were made periodically (every 2 h) using a steady state porometer (Model 1600, Li-Cor, Inc. Lincoln, Nebraska). Two or three leaves on each sapling were labelled with coloured thread so that repeat measurements could be made on the same leaves. This reduced the confounding influence of microclimate on the interpretation of progressive or diurnal changes in stomatal conductance.

The data summarized in the accompanying figures are typical of those observed concurrently on replicate samples. Within each experiment at least three (or as many as five) samples of each species or cultivar were monitored for changes in stem water potential, transpiration and/or stomatal conductance.

RESULTS

The sensitivity of stem water potential to solar radiation was demonstrated by measurements on potted ash trees in the greenhouse (Fig. 1).

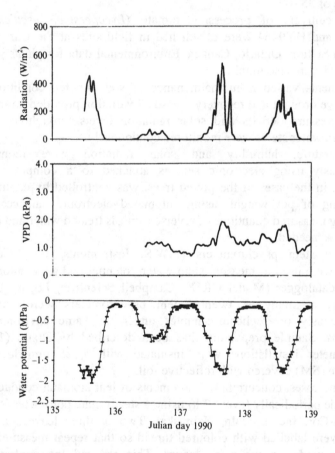

FIG. 1. Changes in the diurnal course of automated stem water potential measurements averaged for three well-watered red ash (*Fraxinus pennsylvanicum*) saplings in pots under greenhouse conditions. During periods of slow changes in water potential (*e.g.* overnight or near midday) the range of measurements was 0.2 MPa but increased to 0.5 MPa during periods of rapid change. Concurrent measurements of total incoming radiation and ambient vapour pressure deficit are shown.

These data represent the mean of stem water potential measurements on three sample trees.

The variability in stem water potential among these samples was 0.2 MPa except during periods of rapid change when it increased to 0.5 MPa. The increased variability was probably due to variability in microclimate.

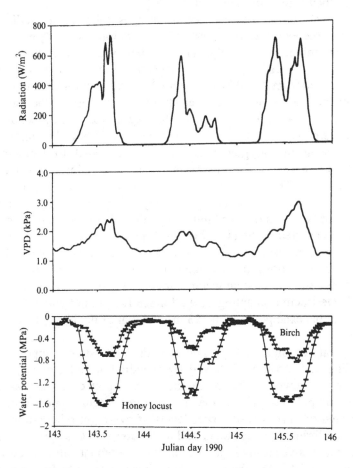

FIG. 2. Diurnal course of automated stem water potential measurements in silver birch (*Betula verrucosa*) and thornless honey locust (*Gleditsia inermis*) potted saplings in a greenhouse environment. Concurrent measurements of total incoming radiation and ambient vapour pressure deficit are shown.

The low level of incoming radiation on day 136 was indicative of cloudy conditions which resulted in relatively moderate (>1.0 MPa) stem water potentials in the trees. Subsequent days were much brighter with a

concurrent doubling of the maximum vapour pressure deficit. These
conditions lead to significant declines in stem water potential relative to
the low radiation day (Fig. 1).

Concurrent environmental data (radiation and vapour pressure deficit)
and stem water potential on representative samples of well watered birch
and honey locust trees are shown in Fig. 2.

The diurnal course of water status changes followed a predictable
pattern with the extremes of stress and rehydration occurring near midday and overnight respectively. The ambient vapour pressure deficit
ranged from 1-3 KPa (being least overnight) while total incoming
radiation ranged from 0-800 W m^{-2}. Note that the radiation level normally
peaked in the late morning, dropped during the middle of the day and
then returned to a high level late in the day unless cloudy conditions
prevailed (*e.g.* day 144). This pattern resulted from the automated
shading system in the greenhouse which responded to temperature
feedback as a means to avoid excessive heat build-up in the greenhouse
during the day. The temperature threshold was set at 27 °C. This
interruption of incoming radiation invariably resulted in a modification to
the level of water stress exhibited by the plants in that a minor
rehydration occurred or the rate of decline in stem water was slowed. The
ambient vapour pressure deficit also responded to the shade control and
the resulting variations in environmental demand for moisture often
mirrored the response of stem water potential in some species (*e.g.* birch,
day 145). Destructive sampling of the foliage found the total leaf area of
the birch in this example to be 0.623 m^2 and that of the honey locust was
0.849 m^2.

A typical example of progressive drought stress imposed by
withholding soil moisture followed by rehydration and recovery upon
rewatering is shown in Fig. 3. In this case a potted red ash sapling
exhibited a progressive decline in daytime stem water potential
accompanied by successively lower levels of overnight rehydration. The
lack of equilibrium between stem and soil water status is apparent from
the lack of a relative steady state in stem water potential values overnight.
In the middle of day 150 the soil moisture was restored to field capacity
resulting in a rehydration phase which culminated with full hydration that
evening. The high and relatively stable stem water potential overnight
indicates virtual equilibrium with the bulk soil water potential. Even
though environmentally demanding conditions prevailed on day 151
(after watering) with both high radiation and vapour pressure deficit, the
stem water potential remained quite high (> -0.5 MPa).

The environmental conditions throughout this period were characterized by mainly cloudless, dry days. The break in the VPD data was the result of sensor failure. The midday drops in radiation were, once again, caused by the automatic shade system responding to high temperature in the greenhouse. The slight perturbations in the slopes of the daily declines in stem water potential are evidence of the sensitivity of the plant's water status to energy balance changes.

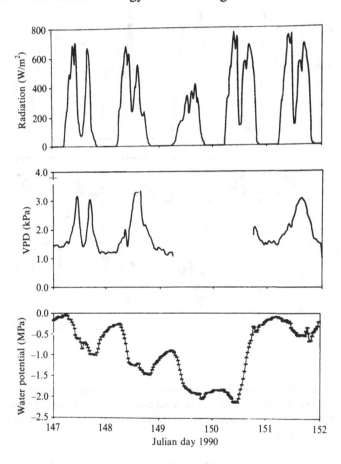

FIG. 3. Diurnal course of automated stem water potential measurements in red ash (*Fraxinus pennsylvanicum*) potted sapling allowed to dehydrate over several days. The root medium was watered to field capacity in the middle of day 150. Concurrent measurements of total incoming radiation and ambient vapour pressure deficit are shown. The gap in VPD data was the result of temporary sensor failure. The midday declines in radiation resulted from the action of automated shades in the greenhouse responding to high temperature.

Diurnal variations in stem water potential and stomatal conductance of two leaves in the canopy of a well watered potted red ash sapling are the subject of Fig. 4.

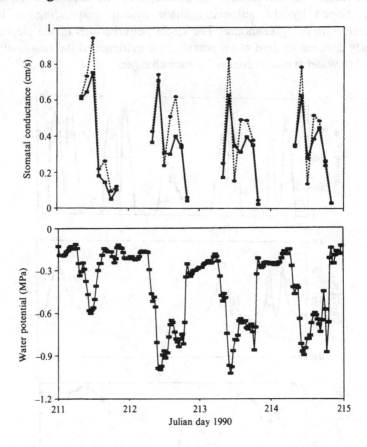

FIG. 4. Diurnal course of automated stem water potential measurements in a red ash (*Fraxinus pennsylvanicum*) potted sapling with concurrent daily measurements of leaf stomatal conductance. Conductance data represent periodic measurements on the same two leaves in the canopy.

The first day (day of year 211) depicts a moderate level of stem water potential (not less than -0.6 MPa) accompanied by a peak in stomatal conductance near midday and a steep decline to virtual stomatal closure by evening. Subsequent days exhibited a "W-function" shaped curve in both water status and stomatal conductance. This period was characterized by environmental data similar to that shown in Fig. 3 (data not shown). Whether influenced by water stress induced closure or the action of the shade system (or a combination of the two), the stomata exhibited partial midday closure followed by an increase in conductance until sun down forced virtual closure. This phenomenon was common among most of the trees growing in the greenhouse, especially on dry days with high radiation loads. The partial stomatal closure resulted in significant midday rehydration as measured by the stem psychrometer. Increases of 0.5 MPa were not uncommon and this level of rehydration often lead to reopening of stomata as shown in Fig. 4.

A similar phenomenon was evident in tomatoes growing in the field (Fig. 5). Concurrent measurements on two cultivars of processing tomatoes showed that under the same environmental conditions, one (LA1563) exhibited the "W-function" response, implying partial stomatal closure during midday water stress, while the other (HY9464) exhibited significantly lower midday stem water potential and apparently less sensitive stomatal response to water stress.

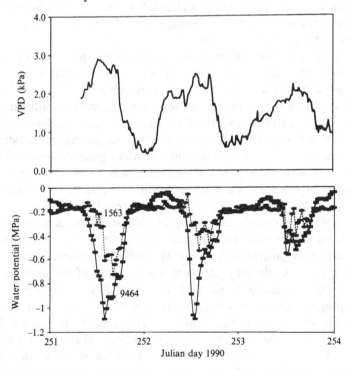

FIG. 5. Automated field measurements of stem water potential in two cultivars of processing tomatoes (*Lycopersicon esculentum*, Cv. LA1563, Cv. HY9464). Concurrent measurements of ambient vapour pressure deficit are shown.

DISCUSSION

Plants routinely undergo diurnal water stress even under well watered conditions. This water stress develops when transpiration exceeds absorption of water by the roots. Plants equilibrate with bulk soil water during rehydration which occurs nightly. Under well watered conditions, daytime water stress was apparent in our experiments even when the bulk soil water potential remained high. The high soil water status was confirmed by the value of the water potential to which the stem rehydrated during the subsequent night without the addition of water to the soil. Thus a non-steady state exists on a diurnal basis in which the water content of the root zone, and to a lesser extent of the stem and leaves of the plant, changes diurnally. The resulting reduction in soil hydraulic conductivity limits the flow of water into the plant root during the day. When transpiration is reduced, sufficient water will flow into the rooting zone increasing both the water potential and conductivity. At no time does the bulk soil water potential fall below the rehydration water potential of the plant on the following night. This further confirms that a localized dehydration must occur.

During progressive soil moisture depletion the relationships among stem water potential, transpiration, vapour pressure deficit and radiation were altered by the stomatal response. Low stem water potentials were prevented by mid-day stomatal closure in some species or cultivars (Figs. 4 & 5). Stem water potentials often increased (became less negative) during stomatal closure. With late afternoon reopening of the stomata, stem water potentials began to decline again but generally reached a less negative potential. The stomatal response was more pronounced and occurred at higher stem water potentials in some species than in others and differences among varieties were also observed.

Although it is outside the scope of this study, it is interesting to speculate on the potential mechanisms of response to water stress in some of the examples presented. The midday partial stomatal closure which contributed to a slight rehydration and re-opening of the stomata ("W-function") was common in many of the plants under high radiation and vapour pressure deficit. It was particularly notable by the absence of this response in one of the tomato cultivars. This resulted in a much lower level of stem water potential in that cultivar (HY9464) relative to LA1563 (Fig. 5). The difference in the level of water stress exhibited by the two species in Fig. 2 also indicates alternative responses to water stress. These differences are not simply due to leaf area (*i.e.* transpiration flux) since the relationship between leaf area and minimum water potential was not clear across species (data not shown). Differences in the conductances of the soil-plant-atmosphere continuum were implicated as important variables in determining the level of water stress exhibited. The potential implications with respect to differences in overall plant water status, turgor and cell expansion, redistribution of water and solutes in the plant and photosynthetic activity are the subjects of ongoing research.

With the introduction of thermodynamic concepts to the field of plant water relations in the early 1960s (Slatyer & Taylor, 1960; Dainty, 1963; Spanner, 1964), many authors attempted to predict the movement of water in the soil-plant-atmosphere continuum (Slatyer, 1967; Cowan, 1965). Slatyer's (1967) interpretation of diurnal variation in water potential in each of these compartments has become a classic and has been reproduced in many texts (*e.g.* Kramer, 1983). In it he predicted that the soil water potential would decline as evapotranspiration removed water from the soil and if water was to move into the plant then its water potential must also decline. In addition to the general decrease in plant and soil water potentials due to evapotranspiration, diurnal variation in plant water potential will also occur when transpiration and absorption rates are unequal. Transpiration can exceed absorption because of the impedance to water movement offered by the soil and the plant itself.

The application of automated and non-destructive means to monitor plant water status has placed us in a position to examine the results of plant-environment interaction with unique temporal resolution. This, coupled with the reliability and accuracy of the *in situ* stem psychrometer, has resulted in these analyses of plant response to water stress and recovery. These empirically demonstrated relationships between plant physiological and environmental variables have permitted unprecedented scrutiny and fine tuning of long held theoretical concepts of plant water relations.

REFERENCES

Coffey, W.L.P. (1989) The morpho-physiological response of three cultivars of potato (*Solanum tuberosum* L.) to varying levels of drought stress. MSc. thesis, University of Guelph, Guelph, Ontario, Canada.

Cowan, I.R. (1965) Transport of water in the soil-plant-atmosphere continuum. *Journal of Applied Ecology*, **2**, 221-239.

Dainty, J. (1963) Water relations of plant cells. *Advanced Botanical Research*, **1**, 279-326.

Darlington, A.S. (1990) Mechanisms of response to atmospheric moisture in greenhouse roses. PhD. thesis, University of Guelph, Guelph, Ontario, Canada.

Dixon, M.A. (1987) *In situ* psychrometry for measuring cut flower water status. *HortScience*, **22**, 285-287.

Dixon, M.A., Grace, J. & Tyree, M.T. (1984) Concurrent measurements of stem density, leaf and stem water potential, stomatal conductance and cavitation on a sapling of *Thuja occidentalis* L. *Plant, Cell and Environment*, **7**, 615-618.

Dixon, M.A. & Tyree, M.T. (1984) A new stem hygrometer, corrected for temperature gradients and calibrated against the pressure bomb. *Plant, Cell and Environment*, **7**, 693-697.

Kramer, P.J. (1983) *Water Relations of Plants*. Academic Press, New York.

McBurney, T. & Costigan, P.A. (1982) Measurement of stem water potential of young plants using a hygrometer attached to the stem. *Journal of Experimental Botany*, **33**, 426-431.

Michel, B.E. (1977) A miniature stem thermocouple hygrometer. *Plant Physiology*, **60**, 645-647.

Neumann, H.H. & Thurtle, G.W. (1972) A Peltier cooled thermocouple dew point hygrometer for *in situ* measurements of water potential. In *Psychrometry in Water Relations* (ed. R.W. Brown and B.P. van Haveren). Utah Agricultural Experiment Station, Utah State University, Logan.

Slatyer, R.O. (1967) *Plant-Water Relationships*. Academic Press, London.

Slatyer, R.O. & Taylor, S.A. (1960) Terminology in plant and soil water relations. *Nature* (London), **187**, 922.

Spanner, D.C. (1951) The Peltier effect and its use in the measurement of suction pressure. *Journal of Experimental Botany*, **2**, 145-168.

Spanner, D.C. (1964) *Introduction to Thermodynamics*. Academic Press, New York.

A proposed mechanism of freezing and thawing in conifer xylem

D.J. ROBSON [1] AND J.A. PETTY [2]

[1] *The BioComposites Centre, University of Wales,
Bangor, Gwynedd, LL57 2UW, U.K.*
[2] *Department of Forestry, University of Aberdeen,
St. Machar Drive, Aberdeen AB9 2UD, U.K.*

ABSTRACT

From the evidence of recent work on freezing in the conifer xylem, previous theories of how cavitation is avoided during freezing and thawing must be rejected. What may happen during freezing is that two pressure rises occur in the tracheid lumen. The first pressure rise occurs as the water expands on crystallising. The second pressure rise is caused by water being drawn out of freezing and unfrozen tracheid lumens, through cell wall capillaries and into the already frozen lumens.

Despite migration of water out of the lumen during freezing, enough water may subsequently be drawn back into the frozen lumen to provide positive pressures during thawing. If positive pressures occur, any bubbles nucleated during freezing quickly re-dissolve. If negative pressures occur in the xylem before the ice has fully thawed then transitory pit aspiration may occur. Pits may deaspirate when pressures in the thawed and thawing tracheids equalise.

INTRODUCTION

The cohesion-tension theory (Dixon & Joly, 1894) is widely accepted as the mechanism of the ascent of sap in tall trees. However, some aspects of this mechanism remain incompletely explained. One of these aspects is the effect of freezing and thawing on the conductivity of conifer xylem.

During freezing gases dissolved in water come out of solution and, at ice/water interface velocities above 2.5 mm s^{-1}, gas bubbles nucleate (Bari & Hallet, 1974). Lybeck (1959), Sucoff (1969) and Robson, McHardy and Petty (1988) observed gas bubbles in frozen xylem tracheids. If bubbles are present in the tracheid lumen when the xylem sap is in tension (*i.e.* at pressures below absolute vacuum) they may expand and cavitate individual tracheids. If more than 25 to 30% of conducting tracheids in a tree cavitate the xylem may become effectively non-conducting (Edwards & Jarvis, 1982; Robson, 1986) (although very limited cell wall flow may still occur). There are two opportunities for cavitation to occur: firstly during freezing as the bubble nucleates at the ice/water interface (Lybeck, 1959); and secondly as the ice surrounding the bubble thaws. Despite these opportunities for cavitation, freezing has not been found to affect conifer xylem conductivity (Hammel, 1967; Sucoff, 1969).

Several theories have been suggested to explain how xylem conductivity is maintained after freezing and thawing (Zimmermann, 1964; Hammel, 1967; Sucoff, 1969, Hygen; 1965).

The Zimmermann theory (Zimmermann, 1964) suggests that tracheids containing gas bubbles cavitate but that these cavitated tracheids are replaced by new growth. This theory only works if very few tracheids contain bubbles. Sucoff (1969) found that 49% of frozen tracheids contained small bubbles and only 8% contained no bubbles, all others appeared to have been previously cavitated. If we assume that all the tracheids which contained small bubbles cavitated, then only 20% of previously full tracheids conduct after each freezing/thawing cycle. Previous work (Edwards & Jarvis, 1982; Robson, 1986) has shown that this reduction in conducting tracheids may make the xylem non-conducting. The Zimmermann theory must be rejected because the xylem will be non-conducting until new water filled tracheids are formed.

Havis (1970) found that conduction continues after the temperature drops below the initial freezing exotherm. Both Hygen (1965) and Havis (1970) conclude that conduction continues in frozen wood by flow through cell wall capillaries. Using this theory cavitation ceases to be a problem because the primary conductive pathways (Havis, 1970) are then the cell wall capillaries. This theory can be simply tested by estimating the flow through cell wall capillaries using the Darcy equation:

$$Q = (K A \Delta P) / (\eta L) \qquad\qquad \text{Equation 1}$$

where Q is the volumetric flow rate, A is the conducting area, ΔP is the pressure difference, η is the viscosity, L is the length of the flow path and K is the permeability constant. If it is assumed that a 10 m tall tree ($L=10$ m), 0.2 m in diameter has a pressure difference between the roots and the crown of approximately 1 MPa, that tracheid walls are 30% of the stem area and that the cell wall permeability (K) is 60×10^{-21} m^2 (Palin & Petty, 1981) then the flow (Q) at 0 °C is 3.14×10^{-14} m^3 s^{-1} or 0.019 ml/week. The Hygen/Havis theory must therefore also be rejected because cell wall flow alone is unable to supply the crown with sufficient water for the tree to survive.

Two plausible but contradictory explanations of how trees survive freezing are suggested by Hammel (1967) and Sucoff (1969). Hammel proposes that the 8.3% increase in volume as water freezes causes a pressure rise in each conducting tracheid. As the pressure rises, pressure differences between freezing and unfrozen tracheids cause the bordered pits to close. It is assumed that water does not migrate out of the freezing tracheid. As the tree thaws the pressure in each thawing tracheid is sufficient to dissolve any bubbles nucleated during freezing. When all the ice in the lumen has thawed the pressures between adjacent tracheids equalise and the bordered pits re-open.

In the Sucoff theory pit aspiration is not required. Instead water flows out of the freezing tracheid into adjacent unfrozen tracheids. The pressure in a freezing tracheid does not increase. In the Sucoff theory the expansion of water on freezing leaves only 92% of the original unfrozen volume of water in the tracheid lumen. On thawing the water and any bubbles in inter-connected lumens are placed in tension. The largest bubble in a group of tracheids expands, and the tracheid containing this bubble cavitates. As water is forced out of this cavitating tracheid the surrounding tracheids are placed under a transient positive pressure. Any remaining bubbles quickly re-dissolve. With the Sucoff theory one conducting tracheid in eleven is cavitated after each freezing and thawing cycle.

In both the Sucoff (1969) and Robson, McHardy and Petty (1988) papers bubbles were observed at the centre of lumens. These observations do not necessarily mean that the water at the centre of the lumen was the last to freeze. Bari and Hallet (1974) reported bands of bubbles in ice. These bands were attributed to the concentration of dissolved gases increasing then being suddenly decreased as a band of bubbles nucleated. A similar process may occur as the ice/water interface moves across the cell lumen.

In recent papers evidence for and against both of these theories of freezing was presented:

1. Pressure rises: in agreement with the Hammel theory, pressure rises of between 0.1 MPa and 3.3 MPa were measured in freezing sapwood blocks (Robson & Petty, 1987);

2. Migration of water: in agreement with the Sucoff theory a theoretical model of freezing in a single tracheid predicted that between 5% and 8% of the water in a tracheid lumen may migrate during freezing (Robson & Petty, 1987). This theoretical calculation was supported by observations

of water droplets exuded by frozen leaf cells (Pearce & Backett, 1985; Jeffree, Read, Smith & Dale, 1987). In the second of these studies (Jeffree, Read, Smith & Dale, 1987) the frozen droplets were found to occur on the cell surfaces facing away from the cryogen, suggesting the presence of a pressure wave in the freezing cell forcing water through the cell wall;

3. Pit aspiration: in agreement with the Sucoff theory both aspirated and unaspirated pits were found in slowly frozen tree stems (Robson, McHardy & Petty, 1988);

4. Wood permeability: in disagreement with the Sucoff theory an 8% reduction in conducting tracheids may significantly reduce xylem conductivity (Edwards & Jarvis, 1982; Robson, 1986). As few as 5 freezing and thawing cycles may make the xylem non-conducting.

In the light of this conflicting evidence neither the Hammel nor Sucoff theories fully explain how the conductivity of trees remains unchanged after freezing and thawing. In addition to the evidence for and against these theories of freezing there is another point which needs more explanation: the theoretical time for wood to freeze (Kubler, 1966; Robson & Petty, 1987) was much shorter than the experimental values (Robson & Petty, 1987).

In this paper we attempt to reconcile these experimental and theoretical results with a mechanism which may explain how trees avoid cavitation during freezing and thawing.

PROPOSED MECHANISM OF FREEZING AND THAWING

Freezing

During freezing the ice/water interface moves from the outer bark to the central axis of branches and stem. As ice starts to form in each tracheid or group of tracheids the pressure rises. Some of the unfrozen water in a freezing tracheid is forced out of the lumen through the tracheid walls. Most bordered pits occur on the radial walls of tracheids. In an ideal cylindrical stem fast flow of water through bordered pits is therefore unlikely because tangentially adjacent tracheids freeze at approximately the same time. Pit aspiration due to the flow of water through the bordered pit membranes or due to pressure differences between adjacent tracheids is therefore also unlikely. However, in a real tree, pit aspiration may occur because of localised variations in structure and density. If the ice/water interface has to pass through the bordered pit membrane

(Robson & Petty, 1987) or if tangentially adjacent tracheids do not freeze at the same time then pit aspiration may occur. It was shown theoretically that up to 8% of the water in each tracheid lumen may migrate through a single tangential wall of a freezing tracheid (Robson & Petty, 1987). Fig. 1 shows three possible routes for water migration from a freezing tracheid. Each of these routes is considered in turn.

At first sight the most likely route for migration is through unfrozen tracheids into already cavitated tracheids towards the centre of the tree (Route 1, Fig. 1). However, migration of water along this route may cause serious problems during thawing. Water which migrates along Route 1 during freezing is still frozen when tracheids nearer the surface of the tree thaw. If there is 5% less water in a fully thawed tracheid then the pressure in the lumen water is approximately -100 MPa. Such low pressures may cause spontaneous cavitation of tracheids. This increased chance of cavitation means that this migration route must be rejected, since such widespread cavitation is not observed.

The flow of water into unfrozen living cells, for example ray parenchyma or the bark and phloem, offers another possible route (Route 2, Fig. 1) for water migration during freezing. One advantage of this route is that each tracheid in the xylem adjoins at least one ray (Esau, 1977). Another advantage is that water stored in living cells during freezing is readily available during thawing. However, this route also has disadvantages: firstly the volume of living cells in the stems of conifers is relatively small, approximately 6% of the total volume (Panshin & De Zeeuw, 1970); secondly, during cold acclimatisation living cells dehydrate in order to prevent ice crystal formation (Levitt, 1980); if water is forced into these cells during freezing they may freeze and die. For these reasons it seems unlikely that water is stored in living cells during freezing.

A third possible route for water migration is through tracheid walls into already frozen tracheids (Route 3, Fig. 1). The water in cell wall capillaries remains unfrozen at temperatures less than the freezing temperature of the lumen water because it is partially bound to the cell wall. However, the main disadvantage of this flow route is that water can only flow through unfrozen capillaries in the tracheid walls. To counteract this increased resistance and the existing pressure difference between freezing and unfrozen tracheids (which causes flow along Route 1) there must be a large pressure difference between the freezing and already frozen tracheids. A possible source of this pressure difference may be the contact of the ice in the tracheid lumen with unfrozen water in cell wall capillaries (Everett, 1961; Dufay, Prigogine, Bellemans & Everett, 1966).

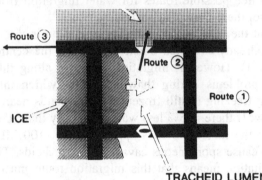

FIG. 1. Diagrammatic representation of freezing in a single tracheid lumen. Three possible routes of water migration are shown: Route 1, flow towards the centre of the tree into unfrozen tracheid lumens; Route 2, flow into living cells in the xylem; and, Route 3 flow away from the centre of the tree into already frozen lumens.

When the lumen is full of ice, growth of the ice crystal into the narrow cell wall capillaries is prevented because any ice formed in capillaries has a higher chemical potential than ice in the lumen (Everett, 1961). The lumen ice can only grow into the capillary ice by increasing its chemical potential. The chemical potential may be increased by increasing the pressure of the lumen ice. If the pressure of the water in the unfrozen capillary remains constant and heat is withdrawn from both the capillary and the frozen lumen then water migrates into the frozen lumen, freezes and increases the pressure of the ice in the lumen until

$$P_S - P_L = (2\,\sigma_{SL}) / R \qquad\qquad \text{Equation 2}$$

where P_S is the pressure of the lumen ice (N m^{-2}), P_L is the pressure of the water in the capillary (N m^{-2}), σ_{SL} is the surface tension of the ice/water interface (approximately 3×10^{-2} N m^{-1}) and R is the radius of the capillary (m). When the pressure difference between the lumen ice and the capillary water is reached the lumen ice crystal grows into the cell wall capillary. This situation is directly analogous to the formation of ice lenses and frost heave in soil. For a median cell wall pore height of 13×10^{-10} m (Stone & Scallan, 1965) the pressure difference ($P_S - P_L$) is approximately 100 MPa. The amount of water drawn into a frozen lumen therefore depends on the compressibility of the tracheid walls, the elasticity of the wood and on the radius of the unfrozen wall capillaries. The pressure difference required to pull water into frozen tracheids may thus be available.

The migration of water along Route 3 during freezing may provide two reasons for the difference between the long freezing exotherm measured experimentally in debarked wood (Robson & Petty, 1987) and the short freezing exotherm calculated theoretically by Robson and Petty (1987) and Kubler (1966). Firstly, by slowing the rate at which heat is drawn from the core of the tree stem: as water freezes it releases latent heat of crystallisation, if more heat is released in the outer part of the stem due to more water freezing then the cooling of the stem core may be slower. Secondly, by reducing the thermal conductivity of the wood: during freezing water may be drawn into frozen tracheids from unfrozen tracheids which are already cavitated or partially cavitated. These empty tracheids then form insulating bands within the tree stem reducing the rate of heat loss from the core of the stem.

The short freezing exotherm calculated theoretically (Robson & Petty, 1987) assumed that the wood only consisted of tracheids. The model therefore slightly overestimated the amount of water that would freeze in a stem by the amount of cell contents in ray cells.

Thawing

While considerable effort has been made to understand the mechanism of freezing, there has been little or no experimental work on the mechanism of thawing. However, two types of thawing are considered below: firstly when the rate of thawing is high and the transpiration rate low; and, secondly when the thawing rate is low and the transpiration rate high.

Low Transpiration Rate - At low transpiration rates there may be time for the ice to thaw and any gas bubbles to dissolve before negative pressures are restored in the xylem. Positive pressures may be maintained in connected tracheids until the last ice melts. Using the equation developed by Epstein and Plesset (1950):

$$Z = R_o^2\, \beta\, /(\, 2\, k\, [c_s - c_i]) \qquad \text{Equation 3}$$

(where Z = time for the bubble to re-dissolve (s), R_o = bubble radius (m), β = bubble gas density (kg m^{-3}), c_s = gas concentration in saturated solution (kg m^{-3}), c_i = gas concentration in liquid (kg m^{-3}), k = coefficient of diffusion of gas into liquid (2×10^{-9} m^2 s^{-1})) the estimated time for the bubbles observed in intact frozen stems (Robson, McHardy & Petty, 1988) to re-dissolve is less than 0.1 s (assuming $c_i /c_s = 0.75$ and $c_s /\beta = 0.02$ (Epstein & Plesset, 1950)). From this calculation it seems clear that any bubbles in the lumen quickly re-dissolve under positive pressures. At low transpiration rates full conduction may thus be restored after all the conducting tracheids thaw.

High Transpiration Rate - When thawing is slow and transpiration rate is high negative pressures may occur in a conducting pathway before thawing is complete. Water may be pulled out of thawed tracheids by transpiration causing negative pressures. Adjacent tracheids which are still thawing may still be at positive pressures due to the presence of ice (Fig. 2a). The pressure difference between thawing and completely thawed tracheids may then cause rapid flow of lumen water and transient pit aspiration. Pit aspiration reduces the rate of flow of lumen water and may give bubbles in the thawing tracheid more time to re-dissolve (Fig. 2b). However, if the pressure in the thawing tracheid drops too quickly and negative pressures develop, the bubbles expand, cavitating the tracheid. Permeability results indicate that cavitation does not occur often.

Assuming that cavitation does not occur, the pressure difference between the tracheids equalises after all the ice has melted due to cell wall flow and the pits deaspirate (Fig. 2c). Pit aspiration and deaspiration may damage the future effectiveness of the pit membrane due to irreversible stretching of the pit margo (membrane) strands. The amount of irreversible aspiration is time-dependent (Thomas & Kringstad, 1971) and so transitory pit aspiration during thawing may have little effect.

DISCUSSION

In brief, the mechanism suggested for freezing and thawing in the conifer xylem without damaging cavitation is:

1. *Freezing.* The pressure of unfrozen water in the lumen of a freezing tracheid rises. Water migrates from the freezing lumen through the tracheid walls. Any bubbles which nucleate at the ice/water interface do not expand because of the positive pressure in the lumen. When the lumen is completely frozen, water is pulled into the frozen lumen out of cell wall capillaries. Flow of water from these capillaries continues until the chemical potential of the lumen ice equals the chemical potential of the capillary water. Water which was forced out of the lumen during

freezing is thus replaced. Unlike in the Hammel (1967) theory pit aspiration need not occur during freezing.

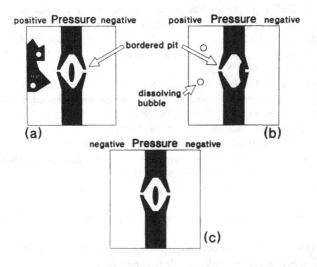

FIG. 2. Diagrammatic representation of the mechanism of thawing at low rates of thawing and high rates of transpiration. (a) A fully thawed tracheid at negative pressure is next to a thawing tracheid at positive pressure. (b) Pit aspiration occurs and bubbles re-dissolve under positive pressures. (c) After thawing is complete pressures equalise and the pit deaspirates.

2. *Thawing.* Two types of thawing are considered: firstly when the rate of thawing is high and the transpiration rate low; and, secondly when the thawing rate is low and the transpiration rate high. For each of these types of thawing a different theoretical mechanism is suggested. When the rate of thawing is high, positive pressures are maintained in the thawing lumen. Bubbles have time to re-dissolve before transpiration causes negative pressures. When the rate of thawing is low adjacent tracheids may be at different pressures. Transient pit aspiration may occur in this situation. If the pressure in isolated thawing tracheids remains positive bubbles will re-dissolve. However, if the pressure becomes negative some cavitation may occur.

ACKNOWLEDGMENTS

We thank the NERC for their financial support of this work.

REFERENCES

Bari, S.A. & Hallet, J. (1974) Nucleation and growth of bubbles at an ice water interface. *Journal of Glaciology*, 8, 489-520.

Dixon, H.H. & Joly, J. (1894) On the ascent of sap. *Annals of Botany*, 8, 468-470.

Dufay, R., Prigogine, I., Bellemans, A. & Everett, D.H. (1966) *Surface Tension and Adsorption*. Wiley, New York.

Edwards, W.R.N. & Jarvis, P.G. (1982) Relations between water content, potential and permeability in stems of conifers. *Plant, Cell and Environment*, 5, 271-277.

Epstein, P.S. & Plesset, M.S. (1950) On the stability of gas bubbles in liquid gas solutions. *Journal of Chemical Physics*, 18, 1505-1509.

Esau, K. (1977) *Anatomy of Seed Plants*. Wiley, New York.

Everett, D.H. (1961) The thermodynamics of frost damage to porous soils. *Transactions of the Faraday Society*, 57, 1541-1551.

Hammel, H.T. (1967) Freezing of xylem sap without cavitation. *Plant Physiology*, 42, 55-66.

Havis, J.R. (1970) Water movement in woody stems during freezing. *Cryobiology*, 8, 581-585.

Hygen, G. (1965) Water stress in conifers during winter. In *Water Stress in Plants* (ed. B. Slavik), pp. 89-95, Dr. Junk Publisher, The Hague.

Jeffree, C.E., Read, N.D., Smith, J.A.C. & Dale, J.E. (1987) Water droplets and ice deposits in leaf intercellular space: redistribution of water during cryofixation for scanning electron microscopy. *Planta*, 72, 20-37.

Kubler, H. (1966) The course of freezing and thawing in logs. Forest Products Laboratory, Ottawa, Canada, 190. Translated from *Holzforschung and Holzverwertung*, 16, 1964, 29-30.

Levitt, J. (1980) *Responses of Plants to Environmental Stresses. Vol. 1 - Chilling, Freezing and High Temperature Stresses*. Academic Press, New York.

Lybeck, B.R. (1959) Winter freezing in relation to the rise of sap in tall trees. *Plant Physiology*, 34, 482-486.

Palin, M.A.& Petty, J.A. (1981) Permeability to water of the cell wall material of spruce heartwood. *Wood Science and Technology*, 15, 161-169.

Panshin, A.J. & De Zeeuw, C. (1970) *Textbook of Wood Technology*. McGraw-Hill, New York.

Pearce, R.S. & Backett, A. (1985) Water droplets in intercellular spaces of barley leaves examined by low temperature scanning electron microscopy. *Planta*, 166, 335-340.

Robson, D.J. (1986) Physical aspects of cavitation and freezing in conifer xylem. PhD Thesis, University of Aberdeen.

Robson, D.J., McHardy, W.J. & Petty, J.A. (1988) Freezing in conifer xylem. II. Pit aspiration and bubble formation. *Journal of Experimental Botany*, 39, 1617-1621.

Robson, D.J. & Petty, Y.A. (1987) Freezing in conifer xylem. I. Pressure changes and growth velocity of ice. *Journal of Experimental Botany*, 38, 1091-1098.

Stone, J.E. & Scallan, A.M. (1965) A study of cell wall structure by nitrogen absorption. *Pulp and Paper Magazine of Canada*, 66, 407-414.

Sucoff, E. (1969) Freezing of conifer xylem and the cohesion tension theory. *Physiologia Plantarum*, **22**, 424-431.

Thomas, R.J. & Kringstad, K.P. (1971) The role of hydrogen bonding in pit aspiration. *Holzforschung*, **25**, 143-149.

Zimmermann, M.H. (1964) The effect of low temperature on ascent of sap in trees. *Plant Physiology*, **39**, 568-572.

Winter xylem embolism and spring recovery in *Betula cordifolia*, *Fagus grandifolia*, *Abies balsamea* and *Picea rubens*

J. SPERRY

Department of Biology, University of Utah,
Salt Lake City, UT 84112, USA.

SUMMARY

Xylem embolism was monitored from mid-winter to mid-summer in four co-occurring species: *Betula cordifolia* (Reg.) Fern., *Fagus grandifolia* Ehrh., *Abies balsamea* (L.) Mill., *Picea rubens* Sarg. The study site was a west-facing slope in the northern Green Mountains of Vermont, U.S.A.; *Betula* and conifers were sampled at 914 m; *Fagus* was collected at 827 m near its local altitudinal limit. Embolism was quantified by the percent the hydraulic conductivity of branch segments was below the maximum obtained following removal of air embolism in xylem conduits. Between early February and early May, the deciduous species averaged 60 to 84% embolism compared to 15 to 60% for the conifers. From April 24 to May 25, embolism in *Betula* dropped from 81 to 8%; this recovery was associated with root pressures up to 86 kPa as measured with manometers at the lower trunk. *Betula* trees in which root pressure was eliminated by overlapping saw cuts still showed 75% embolism in June; only 4% was present in control trees cut in a similar fashion after leaf flush. Root pressure was weak (3 kPa) and uncommon in *Fagus*, and trees remained 80% embolized through June showing considerable dieback. *Fagus* at lower elevations (60 m) were 33% embolized in June with no dieback. Embolism in the conifer species decreased from 40% embolism in late April to 6% in late June despite no detected positive xylem pressures. The mechanism for conifer recovery is unknown.

INTRODUCTION

Xylem embolism occurs as a result of water stress and freezing of xylem sap and it has the immediate consequence of reducing the hydraulic conductivity of the xylem. Longer-term consequences could include reduced growth and dieback. Seasonal studies on deciduous temperate trees and vines have shown that during the winter up to 100% of the hydraulic conductivity can be eliminated by freezing-induced embolism; growing season values may be considerably lower (Sperry *et al.*, 1987; Sperry, Donnelly & Tyree, 1988; Cochard & Tyree, 1990). This suggests xylem is more vulnerable to freezing than to water stress in these species. This may be a result of very different mechanisms for embolism formation by the two stresses.

Experimental evidence indicates embolism caused by water stress can be explained by entry of air through inter-conduit pit membranes (Fig. 1, lower arrows; Crombie, Hipkins & Milburn, 1985; Sperry & Tyree, 1988; Sperry & Tyree, 1990). Perhaps the most compelling observation is that the relationship between embolism and xylem tension in water-stressed stems is identical to that between embolism and air-injection pressure in hydrated stems. Furthermore, the embolism threshold for both air injection and water stress can be varied by treatments altering permeability of inter-conduit pit membranes to air-water menisci (Sperry & Tyree, 1990). If even a few air-filled xylem conduits are present (*e.g.*, because of leaf-abscission) the continuity of water conduction depends on capillary forces preventing passage of an air-water interface between conduits.

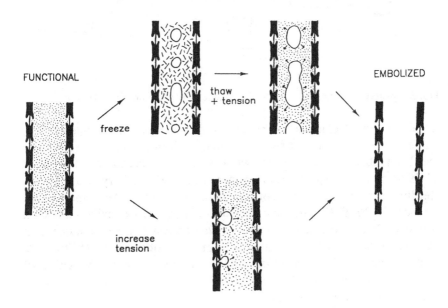

FIG. 1. Embolism formation by freezing (upper arrows) and by water stress (lower). Vulnerability to freezing may be a function of conduit volume; vulnerability to water stress depends on permeability of inter-conduit pit membranes to air-water interfaces.

Freezing-induced embolism can occur because of the low solubility of gases in ice (Fig. 1, upper arrows; Hammel, 1967). Gases dissolved in xylem sap freeze out to form bubbles that potentially nucleate cavitation if they are large enough and if xylem tensions are high enough (Oertli, 1971). Evidence suggests larger-volume conduits are more vulnerable than small ones; presumably because larger bubbles form in higher-volume conduits (Ewers, 1985; Cochard & Tyree, 1990). Conifer tracheids are apparently resistant, although the reason for this is unclear (Hammel, 1967; Sucoff, 1969; Robson, McHardy & Petty, 1988).

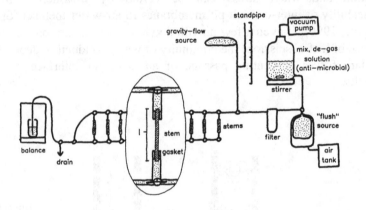

FIG. 2. Apparatus for measuring hydraulic conductivity and embolism. See text.

Given that deciduous temperate trees and vines are vulnerable to winter embolism, it is not surprising that re-filling of embolized vessels has been observed in at least two cases: *Acer saccharum* (Sperry *et al.,*1988) and *Vitis* species (Scholander *et al.,* 1955; Sperry *et al.,* 1987). Refilling was attributed to root and stem pressures in *Acer* and to root pressure in *Vitis*. Circumstantial evidence suggests extensive embolism occurs daily in some herbs because of water stress (*Plantago major, Zea mays*) and that vessels are re-filled at night by root pressures (Milburn & McLaughlin, 1974; Tyree *et al.,* 1986). For these plants, the action of positive pressures in refilling embolized vessels may be critical for their performance and perhaps survival.

The purpose of the present study was to compare winter embolism levels and spring recovery in four co-occurring tree species of the northeastern United States: *Betula cordifolia, Fagus grandifolia, Picea rubens,* and *Abies balsamea*. The genus *Betula* is well known for its ability to generate root pressure in spring and fall (Johnson, 1944), but it has not been demonstrated that these pressures act in re-filling of embolized vessels. In contrast, no report of root pressures in *Fagus grandifolia* was found in the literature. As mentioned, conifers are apparently resistant to freezing-induced embolism, and although they have the ability to generate root pressures under experimental conditions (O' Leary & Kramer, 1965; Lopushinsky, 1980), there seem to be no reports of positive pressures occurring in mature trees in the field.

MATERIALS AND METHODS

The study site was the west side of Camels Hump mountain in northern Vermont. The *Betula* and conifer site was ca. 914 m, the *Fagus* site was ca. 823 m near the foot of the conifer zone, and near the altitudinal limit for this species in northern Vermont. Weather data were collected by a station at the 914 m site operated by Timothy Perkins of the University of Vermont.

Measurement of embolism

Embolism was measured every one to five weeks from early February 1989 to July 1989. For each species, 10 branches from the crown were collected with a pole pruner from the same group of mature trees and brought back to the lab in plastic bags. In the laboratory, one segment 10 to 20 cm long and 0.3 to 1.5 cm in diameter was cut from each of these branches giving ten segments per species. Cuts were made underwater and segments were far enough from the original cut end of the branch to avoid including vessels or tracheids embolised by the initial cut. Segments were fitted with rubber gaskets cut from latex tubing, trimmed with a sharp razor blade, and fitted to a tubing apparatus designed to measure the hydraulic conductivity (Fig. 2).

Hydraulic conductivity was defined as mass flow rate (kg s^{-1}) of measuring solution through a segment divided by the pressure gradient (MPa m^{-1}). The solution used was 10 mol m^{-3} oxalic acid in deionized water; the low pH (ca. 1.8) of this solution inhibited microbial growth. Solution was filtered (220 nm) immediately before use (see Fig. 2). Mass flow rate was measured by routing efflux from a stem segment to a reservoir on a balance; solution flowed into the stem from a source reservoir under a pressure gradient determined by the difference in height between the source and balance reservoirs, and length of stem (Fig. 2).

Typical pressure differences across stems were about 7 kPa. Ten stems at a time were mounted in parallel on the apparatus so conductivity measurements could be made sequentially on each one by directing the flow with stopcocks. After the initial conductivity measurement was completed on all ten segments, flow was routed through all stems at once under a higher pressure of about 100 kPa from a pressurized tank of solution. The purpose of this "flush" was to promote dissolving of air in embolized vessels. Following a flush of 1 to 2 hrs, conductivity of each stem segment was measured again, and the process repeated until conductivity reached a stable maximum value. The percentage by which the initial conductivity was below the maximum gave the "percentage loss in hydraulic conductivity", or a measure of the reversible embolism in each stem segment. This method has been described in detail in Sperry, Donnelly, & Tyree (1987).

Initial measurements on conifers showed declines in conductivity following the flush. This was prevented by de-gassing the solution by agitating it (with magnetic stirrer; Fig. 2) for one hour under vacuum. A better method used in subsequent studies is to draw water into an evacuated container through a jet (e.g., a pipette; Robson pers. commun.). Previous work on conifers (Sperry & Tyree, 1990) has shown that complete recovery of maximum conductivity may be prevented by persistent aspiration of pit membranes. However, this was observed when embolism had been induced by tensions over 7 MPa. These are much higher than typical biological values (less than 2 MPa) for these species. Regardless of whether the torus remained aspirated or not, the maximum conductivities determined for conifers from their native condition would represent the maximum extent of recovery if it occurred.

Stem segments were perfused with 0.05% (w/v) safranin dye to see which xylem conduits were functional. Filtered (220 nm) dye was siphoned through stems for about 30 minutes with 10 kPa sub-ambient pressure.

For *Fagus*, the percentage loss of hydraulic conductivity and the cross-sectional area of the conductivity segment were related to the dry weight of leaves distal to the segment. "Huber values" (Ewers, 1985) were computed as the cross-sectional areas of the stem (estimated from stem diameters) divided by the weight of oven-dried leaves.

Measurement of positive xylem pressures

Manometers were constructed from glass tubes (inner diam. 1 mm) sealed at one end. Tubes were partially vacuum-infiltrated with a weak dye solution, and connected with a minimum of dye-filled tubing to a freshly severed lateral branch near the base of a tree exhibiting positive pressure. Pressure caused by attachment was released by venting the manometer to atmosphere. With the vent closed, the length of the air bubble in the glass tube was measured when it stabilized. The vent was re-opened, and the length of the bubble at atmospheric pressure was recorded. The ratio of bubble lengths gave the root pressure relative to atmospheric assuming bubble length was proportional to volume, and temperature was constant. Results were expressed in kPa above atmospheric (ca. 91 kPa at 914 m).

Before root pressure was observed in spring, overlapping saw-cuts were made on five trees of both *Fagus* and *Betula* in order to minimize root pressures above the cuts. Cuts went halfway into the bole at approximately chest height and were made from opposite sides of the tree, one approximately 0.3 m above the other. A second set of trees for each species was cut in the same way after the leaves had flushed out. Embolism recovery in cut trees was compared to non-cut ones to evaluate the role of root pressure in embolism recovery. The effect of overlapping cuts on root pressure was evaluated with manometers.

RESULTS

Deciduous species

Betula and *Fagus* averaged between 65 and 85% embolism from late January to early May (Fig. 3a). Within each species, embolism did not correlate with segment diameter (0.2 to 1.2 cm), or height above ground (1 to 10 m; data not shown). Fig. 4 shows typical distributions of embolized vessels in transverse sections of branches. In *Betula*, embolism was concentrated in the outer-most xylem of the branch, and was often distributed around the entire circumference. Branches in this species tended to be vertical. In *Fagus,* embolism was also in the outermost xylem, but generally more on the upper side of the branches, most of which were horizontal.

Betula showed an abrupt decrease from 74% embolism on May 1 to 8% on May 25 (Fig. 3a); during this time root pressures were observed and measured. On May 1 no root pressure was seen at the 914 m site, although it was observed at 457 m. By May 8, *Betula* at the site averaged 62% embolism and all trees were showing root pressure. Pressures measured in four trees at noon had maximum values between 15 kPa for a 3 m sapling and 52 kPa on a 7 m tree. The onset of root pressure corresponded with the first above-freezing soil temperatures at 5 cm depth (Fig. 3b). On May 12, *Betula* averaged 35% embolism, and root pressure was still present. Maximum pressures in four trees ranged from 42 to 86 kPa. On May 25, *Betula* was only 8% embolized, buds had broken, and leaves were expanding; root pressure was absent. On June 19, *Betula* was 4% embolized and showing no root pressure.

Betula trees treated with overlapping saw cuts prior to May 1 did not recover from embolism (Fig. 3a, cut), nor did sap flow from wounds made above the cuts. The experiment shown in Fig. 5 demonstrated the effectiveness of overlapping cuts in eliminating root pressure.

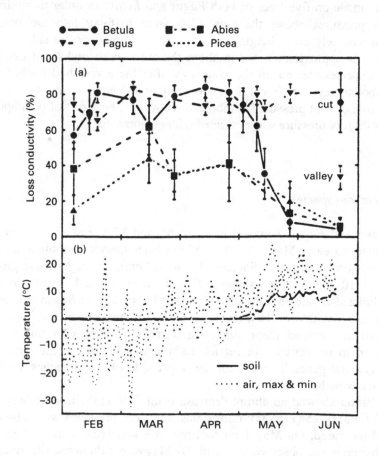

FIG. 3. (a) Embolism (percentage loss hydraulic conductivity) vs. month; means with 95% confidence limits. "Cut": *Betula* with overlapping saw cuts made before May 1. "Valley": *Fagus* from 60 m site; (b) Air and soil (5 cm) temperatures.

Trees given the overlapping cut after leaf flush showed the same embolism levels as non-cut controls.

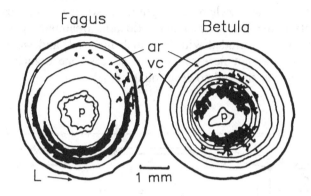

FIG. 4. Transverse sections with dye-stained, functional xylem in black; ar, annual ring; vc, vascular cambium, p, pith. *Fagus*, from June 19 collection; narrow outer ring is current year growth; L, lower side of horizontal branch. *Betula* from a December 12 collection; vertical branch.

While *Betula* was exhibiting consistent root pressure and embolism recovery during May, *Fagus* was showing neither. Embolism remained between 70 and 80%, and root pressures were only occasionally observed. These were weak, measuring up to 3 kPa. Trees with overlapping cuts prior to May 1 had the same embolism levels as non-treated controls and trees given cuts after leafing out (data not shown).

At the June 19 measurement, dieback was observed in *Fagus* at the 823 m site. Leaves were curled and dried on many of the branches of both small and large trees.

This was probably not frost damage because there were no subfreezing temperatures after leafing out in late May (Fig. 3b), and other species in the same locale were healthy. In addition, there was no obvious evidence of a pathogen. Production of new xylem had a minimal impact on conductivity; new xylem was sparse in mid-June (Fig. 4) and was still minimal as late as July 27 when the trees were still 74% embolised (data not shown). *Fagus* in a low elevation forest (60 m) on June 26 was 33% embolized and no dieback was observed (Fig. 3b, valley).

Fagus branches at the site bearing no dead foliage had significantly higher (*t* test, p=0.05) Huber values (1.10 x 10^{-2} m^2 kg^{-1}, n=9) than the low-elevation trees (0.38x10^{-2} n=9) in July. This meant that for a given stem diameter, the low-elevation *Fagus* supported more foliage than the one at the study site. Within the study site there was no correlation between the amount of embolism in a branch and the weight percentage of dead foliage. Branches that bore only dead foliage were either 100% embolized, or their xylem was plugged so the branches had no conductivity either before or after the flushing treatment. The vessels in these stems were occluded with brownish material, and were air-filled. The converse was not necessarily true: at least one branch was 100% embolized and yet showed no dead foliage at all.

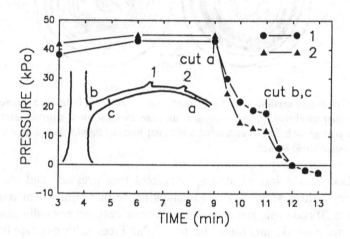

FIG. 5. Xylem pressure (kPa above atmospheric) in *Betula* measured in manometers at branch stubs 1 and 2 (inset). Cutting at a dropped pressures, especially in manometer 2 near the cut. Overlapping cuts at b and c eliminated positive pressures; they became slightly sub-ambient because the cuts were below the manometers.

Conifer species

During the winter, the two conifer species showed considerably less embolism than the deciduous species (Fig. 3a). With the exception of a peak in early March *(Abies, 60%; Picea, 42%)* values did not exceed 40%. The March peak corresponded with an exceptionally cold period just preceeding collection (Fig. 3b). Between the April and June measurements, there was an almost complete recovery from embolism; in both species it declined from 40% to 6%. This recovery preceeded the initiation of new xylem in late May and was not associated with any detectable positive xylem pressures.

DISCUSSION

Significant levels of embolism were present in all species during the winter months. The highest levels were observed in the deciduous species, *Fagus grandifolia* and *Betula cordifolia* (Fig. 3a). The distribution of embolized vessels near the periphery in these species (Fig. 4) is similar to a previous report for *Acer saccharum* where most winter embolism was found in the southern sides of vertical stems (Sperry *et al.*, 1988). This suggests warming by the sun was important in their formation, perhaps via the freeze-thaw mechanism (Fig. 1). In addition, sublimation and water-stress could contribute. Preliminary experiments with *Betula* suggested sublimation was probably a minor cause of embolism when bark was intact (Sperry, unpublished); water stress in deciduous trees in winter may not create xylem tensions high enough to cause significant embolism in the absence of a freeze-thaw event.

Although embolism in the conifer species was less than in the deciduous trees, it was present to a significant degree and increased during the winter (Fig. 3a). If conifers are resistant to embolism caused by freezing and thawing as the literature suggests (Hammel, 1967; Sucoff, 1969; Robson *et al.*, 1988), the observed values may have been a consequence of water stress caused by foliar transpiration and limited uptake from cold and frozen soils. The greater levels exhibited by *Abies balsamea* in February and March (Fig. 3a) could be a reflection of its greater vulnerability to embolism by water stress compared to *Picea rubens* (Sperry & Tyree, 1990). More work is needed to evaluate the mechanism(s) of winter embolism in temperate deciduous trees and conifers.

Recovery of full hydraulic conductivity in *Betula* was associated with root pressures. Not only did root pressure correlate with the abrupt decrease of embolism between May 1 and May 25, but trees in which it was prevented from reaching the crown by overlapping saw cuts showed no recovery from winter levels (Fig. 3a, cut). In addition, *Fagus* showed little or no root pressure and also showed a complete lack of recovery (Fig. 3a). The implication is that root pressure caused recovery by elevating xylem pressures to near or above atmospheric pressure allowing air bubbles in the vessels to dissolve. Although pressures were measured at the bases of trees, in many cases these would have been sufficient to bring crown pressures to positive levels based on the hydrostatic gradient of 10 kPa per meter.

The pattern of recovery in *Betula* resembles that documented for *Acer saccharum* (Sperry *et al.*, 1988) and *Vitis riparia* (Sperry, *et al.*, 1987); both of which showed rather abrupt recovery from winter embolism that was associated with positive xylem pressure. These studies confirm the importance of these pressures for embolism recovery. Positive pressure mechanisms may be critical to the continued success of these species if their growth is sensitive to reduced xylem conductivity.

The fact that *Fagus* did not recover from winter embolism raises the question of whether this influenced its productivity during the growing season. High embolism levels at the high elevation (823 m) site correlated with dieback and high Huber values, and low embolism levels in the valley (60 m) correlated with no dieback and low Huber values. Perhaps embolism influences the elevational range of *Fagus*. However, a much more thorough study would be required to evaluate any cause-and-effect relationship between embolism and dieback, and assess possible pathogen effects. Presumably the trees in the valley exhibited less embolism than higher-elevation ones (Fig. 3a, valley) because of less harsh winter conditions. However, the magnitude of root pressure and consequent recovery may also be a function of habitat in *Fagus*. If Fagaceae in general lack a refilling mechanism, the advantage of the ring-porous condition found in the family (*Castanea, Quercus*) is obvious: recovery occurs by production of relatively large diameter earlywood vessels that supply the emerging foliage (Cochard and Tyree, 1990). There is apparently no restoration of conductivity to previously embolized vessels.

In contrast to the deciduous species, recovery in conifers occurred without detectable positive pressures. Refilling in conifers has been demonstrated before (Waring & Running, 1978), and current monitoring in the Wasatch mountains of Utah is showing similar patterns to that seen in Fig. 3a (Sperry, unpublished). The mechanism of this recovery is unknown. Perhaps it occurs at slightly sub-ambient xylem pressures where the gas in embolized tracheids is under positive pressure due to surface tension. Although the mechanism is unclear, it may be operable only in small-volume conduits such as tracheids, because recovery was not observed in *Fagus* in the absence of positive pressures.

The significance of embolism formation and reversal in woody plants depends on how it influences growth. How redundant is the xylem? Models suggest trees have little redundancy when overall hydraulic architecture is considered (Tyree & Sperry, 1988), but this has not been experimentally demonstrated. It is possible that even fairly extensive embolism would not cause rapid decline in health. Interestingly, the *Betula* that were treated with overlapping saw cuts and did not recover from embolism (Fig. 3a, cut) showed no visual symptoms of decline after one month of being on average 75% embolized. Similarly one *Fagus* branch was found that was 100% embolized and yet supported living foliage. The initial effect of excessive embolism may be a reduction in stomatal conductance followed at some point by leaf senescence and subsequent branch death. The effect of hydraulic conductivity on leaf physiology needs to be quantified before we can evaluate the adaptive advantage of embolism avoidance and recovery mechanisms.

ACKNOWLEDGEMENTS

I thank Timothy Perkins for use of his weather data and research site on Camels Hump.

REFERENCES

Cochard, H. & Tyree, M.T. (1990) Xylem dysfunction in *Quercus:* vessel sizes, tyloses, cavitation and seasonal changes in embolism. *Tree Physiology,* **6,** 393-407.

Crombie, D.S., Hipkins, M.F. & Milburn, J.A. (1985) Gas penetration of pit membranes in the xylem of *Rhodondendron* as the cause of acoustically detectable sap cavitation. *Australian Journal of Plant Physiology,* **12,** 445-453.

Ewers, F.W. (1985) Xylem structure and water conduction in conifer trees, dicot trees, and lianas. *I.A.W.A. Bulletin,* **6,** 309-317.

Hammel, H.T. (1967) Freezing of xylem sap without cavitation. *Plant Physiology,* **42,** 55-66.

Johnson, L.P.V. (1944) Sugar production by white and yellow birches. *Canadian Journal of Research, Section C,* **22,** 1-6.

Lopushinsky, W. (1980) Occurrence of root pressure exudation in Pacific Northwest conifer seedlings. *Forest Science,* **26,** 275-279.

Milburn, J.A. & McLaughlin, M.E. (1974) Studies of cavitation in isolated vascular bundles and whole leaves of *Plantago major* L. *New Phytologist,* **73,** 861-871.

Oertli, J.J. (1971) The stability of water under tension in the xylem. *Zeitschrift fur Pflanzenphysiologie,* **65,** 195-209.

O'Leary, J.W. & Kramer, P.J. (1965) Root pressure in conifers. *Science,* **145,** 284-285.

Robson, D.J., McHardy, W.J. & Petty, J.A. (1988) Freezing in conifer xylem. *Journal of Experimental Botany,* **39,** 1617-1621.

Scholander, P.F., Love, W.E. & Kanwisher, J.W. (1955) The rise of sap in tall grapevines. *Plant Physiology,* **30,** 93-104.

Sperry, J.S., Donnelly, J.R. & Tyree, M.T. (1987) A method for measuring hydraulic conductivity and embolism in xylem. *Plant, Cell and Environment,* **11,** 35-40.

Sperry, J.S., Donnelly, J.R. & Tyree, M.T. (1988). Seasonal occurrence of xylem embolism in sugar maple *(Acer saccharum). American Journal of Botany,* **75,** 1212-1218.

Sperry, J.S., Holbrook, N.M., Zimmermann, M.H. & Tyree, M.T. (1987) Spring filling of xylem vessels in wild grapevine. *Plant Physiology,* **83,** 414-417.

Sperry, J.S. & Tyree, M.T. (1988) Mechanism of water stress-induced xylem embolism. *Plant Physiology,* **88,** 581-587.

Sperry, J.S. & Tyree, M.T. (1990) Water-stress-induced xylem embolism in three species of conifers. *Plant, Cell and Environment,* **13,** 427-436.

Sucoff, E. (1969) Freezing of conifer xylem and the cohesion-tension theory. *Physiologia Plantarum,* **22,** 424-431.

Tyree, M.T., Fiscus, E.L., Wullschleger, S.D. & Dixon, M.A. (1986) Detection of xylem cavitation in corn under field conditions. *Plant Physiology,* **82,** 597-599.

Tyree, M.T. & Sperry, J.S. (1988) Do woody plants operate near the point of catastrophic xylem dysfunction caused by dynamic water stress? Answers from a model. *Plant Physiology,* **88,** 574-580.

Waring, R.H. & Running, S.W. (1978) Sapwood water storage: Its contribution to transportation and effect upon water conductance through the stems of old-growth Douglas fir. *Plant, Cell and Environment,* **1,** 131-140.

Drought resistance strategies and vulnerability to cavitation of some Mediterranean sclerophyllous trees

S. SALLEO AND M.A. LO GULLO

*Istituto di Botanica, Università di Messina,
via P. Castelli 2, 98100 Messina, Italy.*

SUMMARY

The vulnerability to cavitation-induced xylem embolism of three Mediterranean sclerophyllous trees, namely *Ceratonia siliqua* L., *Olea oleaster* Hoffmgg et Link and *Laurus nobilis* L. was investigated as a major factor in their overall drought resistance. Also the capability of the three species of recovering from xylem embolism was studied. *C. siliqua* and *O. oleaster* plants suffered only minor damage to the hydraulic conductivity of their young twigs, under water stress conditions simulating those recorded in the field in May and September in Sicily and in September in Turkey. *L. nobilis* plants, on the contrary, appeared not only to be much more vulnerable to xylem embolism but they were also not capable of repairing damage to xylem water conductivity within 24 h of rewatering. This may explain why leaf water potential remained much more negative than xylem water potential in rewatered Laurel plants, because xylem embolism in leaf petioles prevented water being transported to leaves. The analysis of the distribution of xylem conduit diameter in the leaf petioles of the three species, showed that *L. nobilis* petioles had wider xylem conduits than the other two species. In this comparison, wider xylem conduits appear to be more vulnerable to cavitation than narrower ones. The major vulnerability of *L. nobilis* to cavitation is related to the different distribution of this species with respect to *C. siliqua* and *O. oleaster* within the Mediterranean basin region.

INTRODUCTION

Numerous structural features combine to determine the overall drought resistance of a plant species, including the architecture of the water conducting system (Baas, 1982; Carlquist, 1977; Ewers, 1985; Lo Gullo, Salleo & Rosso, 1986; Salleo & Lo Gullo, 1986, 1989a, 1990; Zimmermann, 1978, 1983). In recent years, much of the debate has concerned the functional importance of xylem conduit features as far as they determine the efficiency of vertical water conduction, and safety under the extreme tension of water stress conditions. In other words, the ideal water conducting system would be that allowing a flow as large as possible at water potentials as low as necessary (Cochard & Tyree, 1990; Lo Gullo, 1989; Lo Gullo & Salleo, 1990; Salleo & Lo Gullo, 1989a, 1989b; Tyree & Dixon, 1986; Tyree & Sperry,1989).

As xylem tensions increase, a threshold is reached at which water starts to cavitate. Xylem cavitation is now recognized as the most common cause of xylem disfunction (Tyree & Sperry, 1989). Therefore, the resistance to cavitation of plant species can be regarded as a major factor in their overall drought resistance.

Previous studies (Lo Gullo & Salleo, 1988) had showed that mediterranean trees like *Ceratonia siliqua* L., *Olea oleaster* Hoffmgg. et Link and *Laurus nobilis* L., even if growing in the same environmental conditions, can develop quite different drought resistance strategies. In particular, *C. siliqua* showed a drought resistance based on maintaining a dynamic equilibrium between water uptake and loss. *O. oleaster* behaved like a drought tolerant species and *L. nobilis* like a drought avoiding water saver (Levitt, 1980). The three species studied (all with a diffuse porous wood) showed different statistical distribution of their xylem conduit diameters. No vessel in *O. oleaster* 1-year-old internodes was wider than 40 μm in diameter while about 20 % of xylem conduits of *L. nobilis* ones were as wide as between 50 and 80 μm in diameter.

There is evidence indicating that the wider a xylem conduit, the more prone it is to cavitation (Salleo & Lo Gullo, 1986, 1989a, 1989b, 1990; Zimmermann, 1983). Tyree & Sperry (1989) have claimed that rather than the xylem conduit diameter, it is the intervessel pit membrane pore diameter that determines the vulnerability to cavitation of a xylem conduit. Since the xylem conduit size is generally stated to have both phylogenetic and adaptive significance, the aim of the present study was to investigate: a) the vulnerability to xylem cavitation of *C. siliqua*, *O. oleaster* and *L. nobilis* as related to their xylem conduit size, and b) the capability of the tree species to recover from cavitation-induced xylem embolism. This last point is of greatest importance for species growing in arid habitats of the Mediterranean Basin region where only about 8 to 10 % of the annual rainfull is distributed between May and October in only a few rainy days.

MATERIALS AND METHODS

Studies were conducted on 5-year-old plants of *C. siliqua*, *O. oleaster* and *L. nobilis* grown in pots at sea level in the Botanical Garden of Messina

(Sicily). Fifty plants of each species were grown in a greenhouse where air temperature and relative humidity varied between 22 and 32 °C and between 30 and 65%, respectively.

All the measurements were performed between April and October 1989. Ten plants of each species were irrigated every 48 h with 200 ml tap water (W-plants) while the other forty plants were divided into four groups of ten plants each which were deprived of irrigation until the desired water stress was reached. These were defined in terms of the ratio of the minimum diurnal leaf water potential (Ψ_{min}) to the leaf water potential at the turgor loss point (Ψ_{tlp}). Different Ψ_{min}/Ψ_{tlp} values were selected simulating field conditions previously recorded by us in May and September in Sicily (*e.g.* Lo Gullo & Salleo, 1988) and by Duhme and Hinckley in Turkey (Duhme & Hinckley, 1992). We will call these plants thereinafter May-S, Sept-S and Sept-T plants, respectively. A higher level of water stress was also applied (that we will call "extreme conditions", E.C.), corresponding to that causing the loss of 10 to 15 % of the leaves within four days after the end of the experiments and after one irrigation (Table 1).

TABLE 1. Levels of the experimental water stress applied, in terms of the ratio of the minimum diurnal leaf water potential (Ψ_{min}) to leaf water potential at the turgor loss point (Ψ_{tlp}). Such ratios simulated those recorded in the field in May and September in Turkey. Extreme conditions corresponded to a more severe water stress, causing the loss of 10-15 % of the leaves within four days after the end of experiments and after one irrigation.

Simulated field conditions	*C. siliqua* Ψ_{min}/Ψ_{tlp} %	*O. oleaster* Ψ_{min}/Ψ_{tlp} %	*L. nobilis* Ψ_{min}/Ψ_{tlp} %
Well-watered	61	60	43
May, Sicily [a]	70	77	73
September, Sicily [a]	91	101	91
September, Turkey [b]	122	121	-
Extreme conditions	131	155	100

[a] Lo Gullo & Salleo, 1988, [b] Duhme & Hinckley, 1989

Water relations parameters

Leaf water potential isotherms were first obtained at 20 °C on five 1-year-old leaves per species, using the pressure bomb technique (Scholander *et al.*, 1965; Tyree & Hammel, 1972). This enabled us to measure Ψ_{tlp}. The procedure employed is described in detail in previous papers (Lo Gullo *et al.*, 1986; Salleo, 1983). The following water relations parameters were measured hourly between 0800 and 1800 h on 1-year-old leaves: a) leaf water potential (Ψ_l); b) xylem water potential (Ψ_x) and c) leaf relative water content (RWC). Ψ_l and RWC were measured as described in detail elsewhere (Lo Gullo & Salleo, 1988; Salleo & Lo Gullo, 1990). Ψ_x was measured on at least three leaves each time by enclosing a number of leaves in plastic bags and wrapping them in aluminium foil on the evening of the day preceding the experiments. This should allow Ψ_l to equilibrate with Ψ_x (Begg & Turner, 1970).

Xylem cavitation and embolism measurements

Xylem cavitation was detected as ultrasound acoustic emissions (AE) (Dixon, Grace & Tyree, 1984; Salleo & Lo Gullo, 1986, 1989a, 1989b) in the frequency range of 100 to 300 kHz. A Drought Stress Monitor (Model 4615, Physical Acoustic Corp., Lawrenceville, USA) was used, equipped with AE transducers (Model R15, Physical Acoustic Corp.) which were clamped to the exposed wood (about 1 cm^2) of 1-year-old internodes. The amplifier was set at a gain of 52 dB.

Xylem embolism was measured in terms of loss of hydraulic conductivity (*L*) in 1-year-old twigs of the tree species (Sperry, Donnely & Tyree, 1987; Sperry, Tyree & Donnelly, 1988).

In particular, the water flow rate (*m*) was measured on twigs detached under water from plants which had reached different levels of water stress (Table 1). From this, *L* was calculated as:

$$L = m / (dP/l) \qquad \text{Equation 1}$$

where *l* is the length of the twig and d*P*/*l* is the pressure gradient which was maintained at 10 kPa. Applying higher pressures (170 kPa) for 20 min and measuring afterwards *m* at a pressure of 10 kPa, the flow rate was found to increase progressively to a maximum. This *m* value was used to calculate the maximum *L*. The "native" hydraulic conductivity was expressed as a percentage of this maximum. All the measurements were performed using 100 mM KCl as the perfusion solution (Salleo & Lo Gullo, unpublished data).

Anatomical measurements

The experimental set-up used to measure the water flow was equipped with a pipette filled with 0.1 % safranin dissolved in 100 mM KCl. The dye was perfused into the twigs for 12 h. Cross-sections of internodes allowed counting of efficient (red-stained) and non-efficient (unstained) xylem conduits as well as measurement of their diameters under a binocular microscope.

Rehydration experiments

Three to five plants of each species, which had been previously water stressed, were rewatered to reach a soil water content of 0.37 g H_2O g^{-1} soil. Eighteen h after the irrigation, they were retested, as described above, for Ψ_l, Ψ_x, RWC and xylem cavitation. Xylem embolism was also measured at 1400 h (24 h after the irrigation). Since Laurel plants showed, in the course of the experiments, no (or very little) recovery from xylem embolism, they were further irrigated every 48 h with 200 ml tap water and tested after 10 days (*i.e.* after five irrigations).

RESULTS

In Figs. 1, 2 and 3, the diurnal time courses of leaf RWC, Ψ_l and Ψ_x are reported for *C. siliqua*, *O. oleaster* and *L. nobilis* plants, respectively, at the different levels of the water stress applied as indicated in Table 1, and after rewatering.

In *C. siliqua* (Fig.1) and *O. oleaster* (Fig. 2) Ψ_l dropped under Ψ_{tlp} (*i.e.* Ψ_l at the turgor loss point) when plants were subjected to water stress corresponding to that suffered by plants in September in Turkey (Sept-T plants) and under what we call "extreme conditions" (E.C. plants).

In Laurel plants (Fig.3), Ψ_l never dropped under Ψ_{tlp} but reached Ψ_{tlp} in E.C. plants.

According to Ψ_l changes, RWC also decreased drastically with the water stress applied, down to 79 % and 72.5 % in plants of *C. siliqua* and *O. oleaster*, respectively. In *L. nobilis* plants, on the contrary, leaf RWC never dropped below 89 %.

In both *C. siliqua* and *O. oleaster,* Ψ_x was maintained only slightly less negative than Ψ_l both in water stressed and in rewatered plants. About 18 h after rewatering, Ψ_l recorded in *C. siliqua* increased significantly over the range of Ψ_l values recorded in irrigated plants (W-plants) and also RWC was about 2 % higher. In contrast, both Ψ_l and RWC recorded in *O. oleaster* rewatered plants increased approximately to the same values recorded in W-plants. In rewatered plants of *L. nobilis* (Fig. 3), big differences were recorded between Ψ_l and Ψ_x . Even though both increased again after rewatering, Ψ_l remained much more negative than Ψ_x , the differences being of the order of 0.3 to 0.95 MPa, respectively. This strongly suggests that a major impediment to water flow was localized between the twig and the leaf xylem (*i.e.* at the leaf traces or in the leaf petiole).

When Laurel plants were retested after having received five irrigations, these differences disappeared both in Sept-T and E.C. plants.

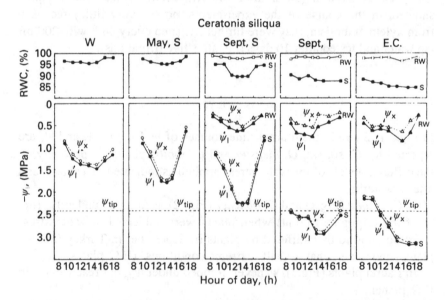

FIG. 1. Time-course of leaf relative water content (RWC), leaf (Ψ_l) and xylem (Ψ_x) water potential. Horizontal dashed lines, labelled as Ψ_{tlp}, yielded the corresponding Ψ_l values at the turgor loss point. W=watered plants; RW=rewatered plants; S=stressed plants; May-S= May in Sicily; Sept-T=September in Turkey, EC= extremeconditions.

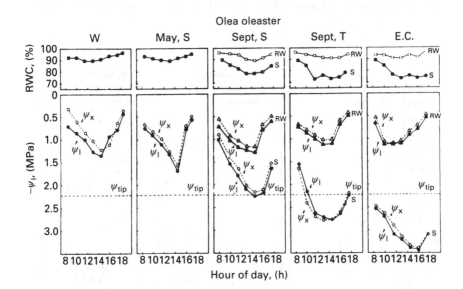

FIG. 2. Symbols and parameters as in Fig. 1.

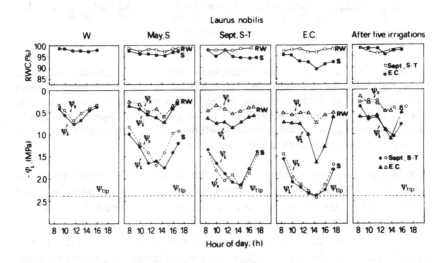

FIG. 3. Symbols and parameters as in Fig. 1.

Xylem cavitation and embolism

In Fig. 4 the numbers of cumulative acoustic emissions (CUM AE) are reported, recorded between 0800 and 1800 h in the three species studied.

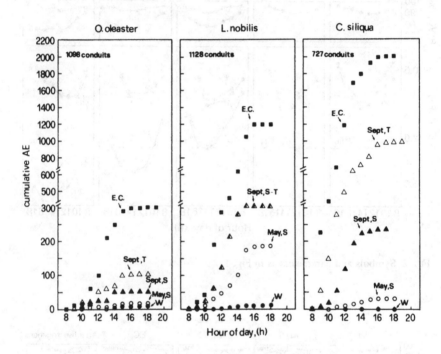

FIG. 4. Time-course of the number of cumulative acoustic emissions (AE) recorded in 1-year-old internodes of watered (W) plants and plants subjected to levels of water stress corresponding to those recorded in the field in May and September in Sicily (May, S and Sept., S) and in September in Turkey (Sept., T). A higher level of water stress was also applied, named "extreme conditions" (E.C.). The number of conduits per section is reported for each species studied (upper left corners).

L. nobilis 1-year-old internodes of May-S and Sept-S plants produced much more AE than those of the other two species. As an example, about 185 and 305 CUM AE were counted in May-S and Sept-S Laurel plants *vs.* 30 and 235 CUM AE in *C. siliqua* analogous plants and only 20 to 50 CUM AE in *O. oleaster* ones. At higher levels of the water stress applied (Sept-T and E.C. plants), *C. siliqua* 1-year-old internodes produced more numerous AE than Laurel ones.

One-year-old twigs of *C. siliqua* Sept-S and Sept-T plant, were showed to suffer only minor damage to their hydraulic conductivity (about 8 %, Fig. 5).

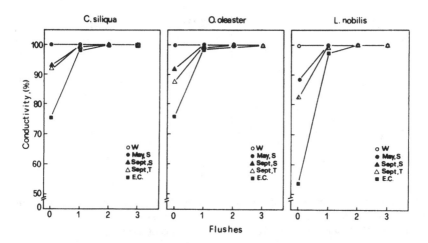

FIG. 5. Changes in the hydraulic conductivity (expressed in % of the maximum) in 1-year-old twigs, before and after applying higher pressures (flushes) to remove xylem emboli. Different symbols indicate watered (W) plants and plants subjected to levels of water stress corresponding to those recorded in the field in May and September in Sicily (May, S and Sept., S) and in September in Turkey (Sept., T). A higher level of water stress was also applied, named "extreme conditions" (E.C.).

One-year-old twigs of *O. oleaster* Sept-T plants lost about 12 % of their hydraulic conductivity (*i.e.* about 50 % more than *C. siliqua* ones).

Laurel plants were the most vulnerable to xylem embolism. The conductivity loss was already about 12 % in May-S plants and increased further to about 18 % in Sept plants and even to 45 % in E.C. ones.

When the conductivity loss (in % of the maximum) was plotted *vs.* the number of CUM AE as recorded between 0800 and 1800 h (Fig. 6), a linear relationship was obtained between the two parameters with fairly high correlation coefficients (*r* ranged from 0.96 to 0.99 in the three species studied). This suggests that the AEs originated in the xylem conduits and to a negligible extent in fibres or other non-conducting wood compartments.

In Fig. 7, the changes in hydraulic conductivity in 1-year-old twigs of watered, water stressed and rewatered plants of the three species are reported *vs.* Ψ_l. Curves were obtained, giving information about the vulnerability of the three species to cavitation-induced xylem embolism as related to a convenient and field-measurable parameter like Ψ_l.

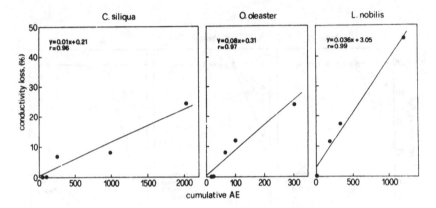

FIG. 6. Loss of hydraulic conductivity in 1-year-old twigs (expressed in % of the maximum) vs. the number of cumulative acoustic emissions (AE) counted between 0800 and 1800 h in plants subjected to increasing levels of water stress.

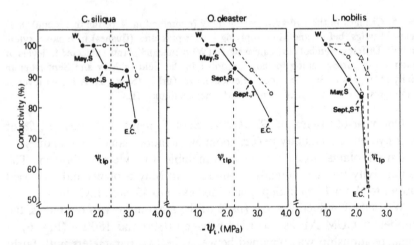

FIG. 7. Hydraulic conductivity (as % of the maximum) of 1-year-old twigs vs. leaf water potential (Ψ_l) of watered (W) plants and of plants subjected to levels of water stress corresponding to those recorded in May and September in Sicily (May, S and Sept., S) and in September in Turkey (Sept., T). A higher level of water stress was also studied, named "extreme conditions" (E.C.). Vertical dashed lines labelled as Ψ_{tlp}, yield corresponding Ψ_l values at the turgor loss point. White circles and triangles indicate analogous conductivity changes after one irrigation (o---o) and five irrigations (Δ-Δ), respectively.

C. *siliqua* plants showed a high resistance to xylem embolism as well as a very good capability of recovering from it. At Ψ_l = -2.1 MPa (as in Sept-S plants) and even at Ψ_l = -3.1 MPa (as in Sept-T plants) *i.e.* at Ψ_l values near to or even more negative than Ψ_{tlp} (Ψ_{tlp} = -2.4 MPa), the hydraulic conductivity loss of 1-year-old twigs was only 8 to 10 %. Under these conditions, the conductivity was completely recovered 24 h after irrigation.

O. *oleaster* plants were somewhat less efficient in recovering from xylem embolism. Even in this species, however, the conductivity loss was only about 8 % in Sept-S and 12 % in Sept-T plants, although Ψ_l values were as negative as -2.9 MPa (*i.e.* more negative than Ψ_{tlp}). Twenty-four h after irrigation, a residual damage to hydraulic conductivity of 1-year-old twigs of *Olea* was recorded, of about 8 % in Sept-T plants.This means that only about one third of the damage was repaired by plants under these experimental conditions.

Laurel plants showed the higher vulnerability to xylem embolism at the least negative Ψ_l values. When Ψ_l approached Ψ_{tlp} (Ψ_l = -2.2 MPa, Ψ_{tlp} = -2.4 MPa), the conductivity loss was about 18 % (Sept-S, T plants) and at Ψ_l = Ψ_{tlp} (E.C. plants), the conductivity loss was as high as 45 %. Moreover, Sept-S, T plants failed to recover from xylem embolism 24 h after the first irrigation and even in May-S plants the recovery was not complete. When Laurel plants were retested after five irrigations (ten days), a much better recovery was recorded. The residual damage to conductivity was only 5 % in Sept-S, T plants and of only 10 % in E.C. plants *i.e.* over 70 % of the damage was repaired.

Anatomy of the leaf petiole

The distributions of xylem conduit diameters in leaf petioles of the three species studied are reported in Fig. 8.

It can be seen that O. *oleaster* petioles had the narrowest conduits (less than 15 μm in diameter). L. *nobilis*, on the contrary, showed the widest conduits (about 18 % of them were between 20 and 25 μm in diameter *vs.* about 9 % in C. *siliqua*). In terms of water flow, this makes a big difference. If water flow is set as proportional to $\Sigma \pi r^4$ where r is the xylem conduit radius, the theoretical flow across L. *nobilis* petioles would be about 10 times greater than that across C. *siliqua* petioles and 18 times greater than that across those of O. *oleaster*.

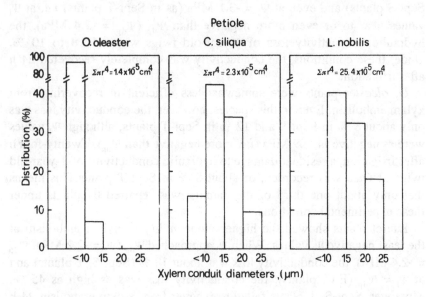

FIG. 8. Xylem conduit diameter distributions in leaf petioles. $\Sigma \pi r^4$ values are also reported, where r is the xylem conduit radius.

DISCUSSION

Among the three species studied, *L. nobilis* was the most vulnerable species to xylem cavitation. Within the "leaf turgor region", (*i.e.* at $\Psi_l >$ Ψ_{tlp}), the young twigs of this species underwent the highest damage to their hydraulic conductivity. Moreover, no short-term recovery from xylem embolism was recorded in them.

C. siliqua, on the contrary, underwent only minor cavitation-induced xylem embolism and showed the most efficient short-term recovery even in plants with Ψ_l values more negative than Ψ_{tlp} ones.

O. oleaster was less efficient than *C. siliqua* in the short-term recovery from xylem embolism but the species can still be regarded as fairly well resistant to xylem cavitation. A loss of hydraulic conductivity of 10 to 12 % is really not much, because of excess functional wood in trees.

After rewatering, Ψ_l of pre-stressed *L. nobilis* plants tended to remain much more negative than Ψ_x (Fig. 3). In other words, water was not efficiently transported into leaves. This strongly suggests that xylem conduits connecting twigs to leaves were embolized. The leaf traces and petioles are points of major resistance to water flow in plants (Isebrands & Larson, 1977). The water potential drop across such regions is steeper than elsewhere. Therefore, the risk of the water cavitation is here higher than in the stem.

Laurel plants showed wider xylem conduits in petioles than the other two species studied. The theoretical flow in Laurel petioles was many times greater than in *C. siliqua* and in those of *O. oleaster*. Zimmermann (1983) hypothesized that the wider is a xylem conduit, the more prone to cavitation it should be. There is a consistent body of evidence that this is the case in different trees like *Chorisia insignis* H. B. et K. (Salleo & Lo Gullo, 1986), *Vitis vinifera* L. (Salleo & Lo Gullo, 1989a) *C. siliqua* L. (Lo Gullo & Salleo, 1989b).

The xylem conduit embolism in Laurel petioles at Ψ_l values less negative than in the other two species studied was probably due to their wider xylem conduit diameters. It is also possible, however, that xylem conduits in *L. nobilis* petioles have wider pit membrane pores than those of *C. siliqua* and *O. oleaster* and that, for this reason, they are more vulnerable to cavitation, as suggested by Tyree & Sperry (1989).

C. siliqua and *O. oleaster* grow on arid hills between 0 and 500 m above sea level throughout the Mediterranean Basin region. The two species are the most typical components of a plant community aptly termed *Oleo-Ceratonion* (Pignatti, 1982). Both have to overcome not only the seasonal drought (usually lasting between April and October) but also the short-term water stress due to the noticeable increase in air temperature and the decrease in air relative humidity during the warmest hours of the day. In this study, both *C. siliqua* and *O. oleaster* have been shown to be very resistant to xylem cavitation, capable of recovering very well and in a short time from xylem embolism, after rewatering.

L. nobilis can also be found as a component of the *Oleo-Ceratonion* community but, in this case, the species tends to reduce its size and to change its typical life form from tree to shrub. This is consistent with its higher vulnerability to cavitation as well as its low capability of recovering from xylem embolism after a brief rain (one irrigation). After a more prolonged humid period (five irrigations), however, *L. nobilis* plants rehydrated very well and the damage to xylem hydraulic conductivity was repaired. In other words, *L. nobilis* seems to be well adapted to resist drought, provided the dry period is interrupted by humid ones; otherwise, the species loses part of its leaves and must, therefore, invest part of its metabolic energy to produce new ones.

In this case, the plant size is reduced and the loss of the youngest peripheral parts of the crown (compensated by the production of more and more proximal shoots), determines the typical change in the life form of the species.

REFERENCES

Baas, P. (1982) Systematic, and ecological wood anatomy - history and perspectives. In *New Perspectives in Wood Anatomy* (ed. P. Baas), pp. 23-58. Martinus Nijhoff, The Hague.

Begg, J. E. & Turner, N.C. (1970) Water potential gradients in field tobacco. *Plant Physiology*, 46, 343-345.

Carlquist, S. (1977) Ecological factors in wood evolution: a floristic approach. *American Journal of Botany*, 64, 887-896.

Cochard, H. & Tyree, M.T. (1990) Xylem disfunction in *Quercus*: vessel sizes, tyloses, cavitation and seasonal changes in embolism. *Tree Physiology*, 6, 397-407.

Dixon, M.A., Grace, J. & Tyree, M.T. (1984) Concurrent measurements of stem density, leaf and stem water potential, stomatal conductance and cavitation on a sampling of *Thuja occidentalis* L. *Plant, Cell and Environment*, 7, 615-618.

Duhme, F. & Hinckley, T.M. (1992) Daily and seasonal variation in water relations of macchia shrubs and trees of France (Montpellier) and Turkey (Antalya).*Vegetatio*, 99-100, 185-198.

Ewers, F.W. (1985) Xylem structure and water conduction in conifer trees, dicot trees, and lianas. *IAWA Bullettin* (n. s.) 6, 309-317.

Isebrands, J.G. & Larson, P.R. (1977) Vascular anatomy of the nodal region in *Populus deltoides* Bartr. *American Journal of Botany*, 64, 1066-1077.

Levitt, J. (1980) *Responses of Plants to Environmental Stresses*. Academic Press, New York.

Lo Gullo, M.A. (1989) Xylem architecture as the anatomical basis of drought resistance in the desert shrub *Simmondsia chinensis* (Link) Schneider. *Giornale Botanico Italiano*, 123, 255-267.

Lo Gullo, M.A. & Salleo, S. (1988) Different strategies of drought resistance in three Mediterranean sclerophyllous trees growing in the same environmental conditions. *New Phytologist*, 108, 267-276.

Lo Gullo, M.A. & Salleo, S. (1990) Wood anatomy of some trees with diffuse- and ring-porous wood: some functional and ecological interpretations. *Giornale Botanico Italiano*, 124.

Lo Gullo, M.A., Salleo, S. & Rosso, R. (1986) Drought avoidance strategy in *Ceratonia siliqua* L., a mesomorphic-leaved tree in the xeric Mediterranean area. *Annals of Botany*, 58, 745-756.

Pignatti, S. (1982) *Flora d'Italia*. Edagricole, Bologna.

Salleo, S. (1983) Water relations of two Sicilian species of *Senecio* (Groundsel) measured by the pressure bomb technique. *New Phytologist*, **95**, 179-188.

Salleo, S. & Lo Gullo, M.A. (1986) Xylem cavitation in nodes and internodes of whole *Chorisia insignis* H.B. et K. plants subjected to water stress: relations between xylem conduit size and cavitation. *Annals of Botany*, **58**, 431-441.

Salleo, S. & Lo Gullo, M.A. (1989a) Xylem cavitation in nodes and internodes of *Vitis vinifera* L. plants subjected to water stress. Limits of restoration of water conduction in cavitated xylem conduits. In *Structural and Functional Responses to Environmental Stresses: Water Shortage* (ed. K.H. Kreeb, H. Richter & T.M. Hinckley), pp. 33-42, SPB Academic Publishing, The Hague, The Netherlands.

Salleo, S. & Lo Gullo, M.A. (1989b) Different aspects of cavitation resistance in *Ceratonia siliqua* a drought-avoiding Mediterranean tree. *Annals of Botany*, **64**, 325-336.

Salleo, S. & Lo Gullo, M.A. (1990) Sclerophylly and plant water relations in three Mediterranean *Quercus* species. *Annals of Botany*, **65**, 259-270.

Scholander, P.F., Hammel, H.T., Bradstreet, E.D. & Hemmingsen, E.A. (1965) Sap pressure in vascular plants. *Science*, **148**, 339-346.

Sperry, J.S., Donnelly, J.R. & Tyree, M.T. (1987) A method for measuring hydraulic conductivity and embolism in xylem. *Plant, Cell and Environment*, **11**, 35-40.

Sperry, J.S. & Tyree, M.T. (1988) Mechanism of water stress-induced xylem embolism. *Plant Physiology*, **88**, 581-587.

Sperry, J.S., Tyree, M.T. & Donnelly, J.R. (1988) Vulnerability of xylem to embolism in a mangrove vs. an inland species of *Rhizophoracea*. *Physiologia Plantarum*, **74**, 274-283.

Tyree, M.T. & Dixon, M.A. (1986) Water stress induced cavitation and embolism in some woody plants. *Physiologia plantarum*, **66**, 397-405.

Tyree, M.T. & Hammel, H.T. (1972) The measurement of the turgor pressure and water relations of plants by pressure-bomb technique. *Journal of Experimental Botany*, **23**, 267-282.

Tyree, M.T. & Sperry, J.A. (1989) Vulnerability of xylem to cavitation and embolism. *Annual Review of Plant Physiology Molecular Biology*, **40**, 19-38.

Zimmermann, M.H. (1978) Hydraulic architecture of some diffuse-porous trees. *Canadian Journal of Botany*, **56**, 2286-2295.

Zimmermann, M.H. (1983) *Xylem Structure and the Ascent of Sap*. Springer-Verlag, Berlin.

Relations between sap velocity and cavitation in broad-leaved trees

M. BORGHETTI,[1,4] P. DE ANGELIS,[2] A. RASCHI,[3]
G.E. SCARASCIA MUGNOZZA,[2] R. TOGNETTI[1] AND R. VALENTINI [2]

[1] *Istituto Miglioramento Genetico delle Piante Forestali,*
Consiglio Nazionale delle Ricerche, via S. Bonaventura 13, 50145 Firenze, Italy.
[2] *Dipartimento Scienze dell'Ambiente Forestale e delle sue Risorse,*
Università della Tuscia, via S. Camillo De Lellis, 01100 Viterbo, Italy.
[3] *Istituto di Analisi Ambientale e Telerilevamento applicati all'Agricoltura, Consiglio*
Nazionale delle Ricerche, p.le delle Cascine 18, 50144 Firenze, Italy.
[4] *Present address: Dipartimento di Produzione Vegetale,*
Università della Basilicata, Potenza, Italy.

SUMMARY

Concurrent measurements of cavitation by the ultrasound acoustic emission technique and sap velocity by the thermoelectric heat pulse method were carried out in the field on three woody species (*Quercus pubescens, Quercus ilex* and *Alnus cordata*) characterized by different wood structure. The plant water status was assessed by measuring xylem water potential and stomatal conductance. A good correspondence was found between the patterns of sap velocity and cavitation rate. A threshold-type relationship was observed, in *Alnus cordata*, between water flow and cavitation rate. In some cases temporal lags between ultrasound emission and sap velocity were observed: several factors may account for these lags, including the possibility that cavitation of xylem conduits may be a rather patchy phenomenon and that different xylematic volumes might have been sensed by the ultrasound and heat pulse transducers.

INTRODUCTION

The formation and spreading of gaseous emboli through the xylem are recognized as common events in water stressed plants (Milburn, 1979; Tyree & Sperry, 1989a). In particular, it is widely held that the increase of xylematic tension, which is caused by the drop of water potential between the soil and the atmosphere, frequently induces cavitation, *i.e.* the breakage of water columns and the formation of gas bubbles in the lumina of xylem conduits.

The mechanism of cavitation is still being debated, although the hypothesis that cavitation is caused by the aspiration of air bubbles through the intervessel pit membranes (the so-called air seeding hypothesis) is widely supported by experimental evidence (Sperry & Tyree, 1988; Sperry, Tyree & Donnelly, 1988).

The embolization of xylem conduits can be regarded as a physiological disfunction, which reduces the hydraulic conductivity and impairs the effectiveness of water transport. Recently, the hypothesis has been put forward that the interactions between the hydraulic conductivity, the tension in xylem sap and the vessel embolization may bring woody plants near a point of catastrophic xylem disfunction (Tyree & Sperry, 1988; Tyree, 1989).

The opportunity for studying embolization processes and water flow through the xylem, and looking into their relationships, benefits from the development of non-destructive techniques, such as the acoustic sensing of cavitation (Milburn & Johnson, 1966; Tyree & Dixon, 1983; Sandford & Grace, 1985) and the thermoelectric methods for measuring xylem flows (cf. Pearcy, Schulze & Zimmermann, 1989 and the papers by Cohen, Fuchs & Moreshet, Cermák, n & Kucera, Valancogne & Nasr, in this volume). Indeed, simultaneous measurements of cavitation and sap velocity may result in a better understanding of functionality and vulnerability of the xylem.

This work was undertaken with the aim of exploring, at the whole plant level and in field conditions, the relations between sap velocity and embolism in broad-leaved trees with different features: *Quercus pubescens* L. (deciduous species characterized by a ring-porous xylem with vessels up to 500 μm in diameter), *Quercus ilex* L. (evergreen species with a diffuse-porous xylem and vessels up to 150 μm in diameter), *Alnus cordata* Loisel. (deciduous species with a diffuse-porous xylem and vessels up to 40 μm in diameter). Knowledge of these relationships may contribute to an understanding of the interactions between xylem efficiency and water stress, which in Mediterranean regions can determine plant survival and productivity.

MATERIALS AND METHODS

This research was carried out at two sites in Central Italy: i) a Mediterranean "macchia" community along the Thyrrenian coast on a coastal sand dune (Lat. 42° 22' N, Long. 11° 32' E), where one *Q. ilex* and one *Q. pubescens* tree (stem diameter at the height of 1.3 m = 11 cm) were studied on June 14 and 15 1989; ii) a plantation growing on a silty-clay soil on the grounds of the Forestry Faculty in Firenze (Lat. 43° 47'

N, Long. 11° 40' E, altitude 40 m) where one *A. cordata* tree (stem diameter at 1.3 m = 10 cm) was studied on July 5 and 6 1989 and on 8 to 11 October 1990.

Ultrasound acoustic emissions (UAE) from the xylem were recorded using an ultrasonic transducer (PAC I151, Physical Acoustics Corp., Princeton, NJ, USA) and amplified 75 dB by a UAE counter (model 4615 Drought Stress Monitor, Physical Acoustics Corp.). A portion of the bark was removed to expose a small area of xylem, on which the transducer was clamped; this area was coated with silicone grease to prevent evaporation from the tissue. The contact between the transducer and the xylem was improved by using an ultrasound transmitting gel.

Xylem flow was measured by the thermoelectric "heat pulse method", near the point where the UAE transducer was clamped. In 1989 a "home-made" measurement set was made up with one heating device inserted into the xylem at a depth of 20 mm and by one thermistor probe (2 mm in diameter) inserted at 15 mm of depth, 15 mm above the heat source. Heat pulse velocity (HPV) was calculated taking account of the time needed to reach a temperature maximum at this fixed distance. In this case one HPV measurement set and one UAE transducer were applied, close to each other, on the stem of the trees, at a height of 1 m. In 1990 commercial HPV equipment (Soil Conservation Centre, Aokautere, New Zealand) was used. The measurement set was made up with four heating devices inserted at different depths into the xylem (from 1 to 4 cm) in four positions at right angles on the stem circumference, and by two thermistor probes (1.8 mm in diameter), at each of the four positions, that were inserted into the xylem 10 mm above and 5 mm below each heating device. In this case The Hüber & Schmidt (1937) compensation technique is used to produce heat pulse velocity, Marshall's theory (1958) and the analysis by Swanson & Whitfield (1981) are used to convert heat to sap velocities, taking account of inhomogeneities caused by sensor implantation wounds. In this case, four UAE transducers were clamped to the stem close to the HPV probes; another two UAE transducers were clamped to major branches and two more to major roots. In both years, sampling intervals ranged from 15 to 60 minutes for HPV and from 2 to 4 minutes for UAE measurements.

The twig xylem water potential and the leaf stomatal conductance were measured at 2 to 3 hour intervals from sunrise to sunset, using a pressure chamber and a null-balance steady state porometer (Li-1600, Li-Cor, Lincoln, Nebraska, USA), respectively. At least three twigs or leaves, sampled in the upper part of the crown, were measured each time.

Air temperature and humidity and shortwave solar radiation were measured, at regular intervals, at the experimental sites.

In order to analyze the relationships between flow and embolism, coinciding series of HPV and UAE values were produced and their cross-correlation function was computed. This analysis involved the calculation of correlation coefficients (r_{xy}) between the two series of data not only for measurements taken at the same time but also at time k apart, where k is the lag. The value of k corresponding to the maximum r_{xy} gives an estimate of the time lag betwen the temporal series of the two variables. All calculations were performed using the StatGraphic statistical package (STSC Company, 1985).

RESULTS

Experiments were conducted on *Q. pubescens* and *Q. ilex* on June 14 and 15 1989, respectively. Both days were warm and sunny, with day-time air temperatures between 10 and 28 °C, shortwave solar radiation up to 1000 W m^{-2} and vapour pressure deficit up to 2.7 kPa.

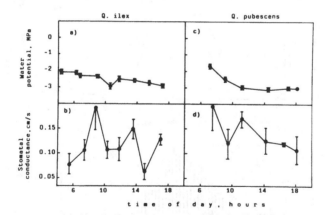

FIG. 1. Daily behaviour of xylem water potential and stomatal conductance in *Q. ilex* (a and b) and *Q. pubescens* (c and d); measurements carried out on June 14 1989 on *Q. ilex* and on June 15 on *Q. pubescens*.

Q. ilex experienced a rather marked water stress. Indeed, the pre-dawn water potential was -2.1 MPa and potentials constantly lower than -2.0 MPa were measured throughout the day, with a minimum of -2.9 MPa at 11:00 h (Fig. 1a). Stomatal conductance (Fig. 1b) displayed a maximum of 0.15 s cm^{-1} at 09:00 h and a midday depression in correspondance to the minimum plant water potential. A sharp and temporary decrease of stomatal conductance in the early afternoon (Fig. 1b) may be attributed to a temporary decrease of solar radiation, which at this time of day was as low as 170 W m^{-2} because of a cloud patch.

The water potential of *Q. pubescens* decreased from -1.6 MPa at 07:00 h to a minimum of -3.0 MPa at 14:00 h (Fig. 1c). The stomatal conductance was higher than 0.1 cm s^{-1} throughout the day, with a maximum of 0.2 cm s^{-1} at 08:30 h and a minimum of 0.11 cm s^{-1} at 19:00 h (Fig. 1d).

In Fig. 2a the daily courses of heat pulse velocity (HPV) and ultrasound acoustic emissions (UAE) in *Q. ilex* are displayed. UAE started to rise at 05:00 h and rose regularly, except for a short reduction between 12:00 and 13:00 h, until 14:30 h when the maximum rate of 6 ct min^{-1} was reached. HPV started to rise approximately two hours after UAE and increased regularly until 16:00 h, when the maximum value of 1.2 m h^{-1} was recorded.

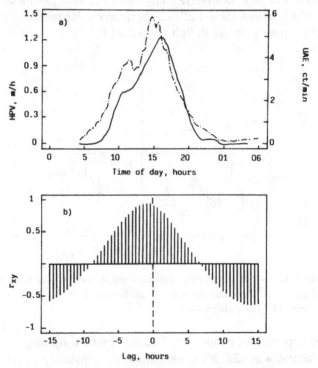

FIG. 2. Daily behaviour of ultrasound acoustic emission (UAE, broken line) and heat pulse velocity (HPV, continuous line) measured in *Q. ilex* on June 14 (a) and the cross-correlation function between their temporal series (b).

The HPV rise was somewhat lower between 10:30 and 13:30 h, nearly two hours before the UAE reduction. Both UAE and HPV showed a similar decreasing trend during the afternoon, reaching steady low values at midnight. The cross-correlation function (Fig. 2b) indicates that the values of UAE and HPV are well correlated ($r > 0.9$) and that there is a maximum lag of one hour between the two time series.

In Fig. 3a, the daily courses of UAE and HPV in *Q. pubescens* are reported. Both UAE and HPV started to rise at 8:00 h; HPV increased regularly until 12:30 h, whereas the rise of UAE was more rapid between 10:30 and 12:00 h, when a maximum of 34 ct min^{-1} was reached. A similar decreasing trend for UAE and HPV was observed during the afternoon. Even in this case the correlation between the two variables was high ($r > 0.9$) and no lag was observed between their temporal series (Fig. 3b).

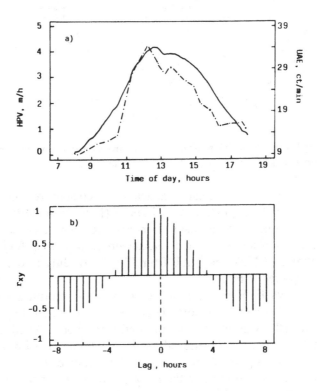

FIG. 3. Daily behaviour of ultrasound acoustic emissions (UAE, broken line) and heat pulse velocity (HPV, continuous line) measured in *Q. pubescens* on June 15 (a) and the cross-correlation function between their temporal series (b).

In 1989 the measurements on *A. cordata* were carried out on two bright sunny days (July 4 and 5), which followed a few rainy days. The air temperature was never lower than 16 °C and went up to 30 °C, the shortwave solar radiation was up to 950 W m⁻² and the vapour pressure deficit up to 2.5 kPa.

During both days the xylem water potential decreased regularly during the morning, from a high dawn value, which on July 5 was -0.25 MPa, to values lower than -1.7 MPa in the afternoon (Figs. 4a and 4b).

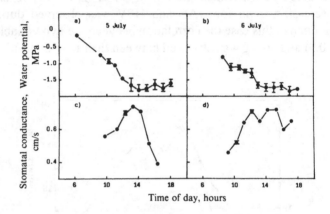

FIG. 4. Daily behaviour of xylem water potential and stomatal conductance measured on July 5 1989 (a and c) and 6 (b and d), in *A. cordata*.

On both days the stomatal conductance increased regularly in the morning, up to values greater than 0.7 cm s⁻¹ (Figs. 4c and 4d). On July 5 a marked decrease of stomatal conductance was recorded in the afternoon, whereas more scattered values were measured on July 6.

The patterns of HPV and UAE from 00:00 h on July 5 to 15:00 h on July 6 are reported in Fig. 5a.

On both days HPV rose rapidly throughout the morning and values greater than 0.3 m h⁻¹ were recorded over several hours. On July 5 a rapid decrease of HPV was evident in the early afternoon. At night, rather variable HPV values were observed: between 0.05 and 0.15 m h⁻¹ during the first night, and up to 0.22 m h⁻¹ during the second one.

On July 5 the rise of UAE in the morning, which reach a peak of 37 ct min⁻¹ at 08:30 h, was shifted by approximately two hours before the rise of HPV. Decreasing values of UAE were recorded in the afternoon and two UAE minor peaks (up to 14 ct min⁻¹) were observed during the night between July 5 and 6. On the morning of July 6 low UAE rates were measured.

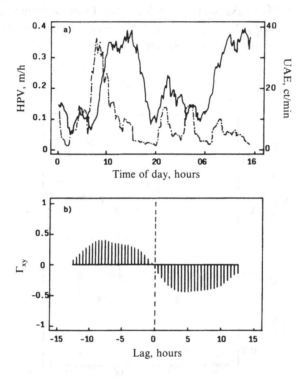

FIG. 5. Daily behaviour of ultrasound acoustic emissions (UAE, broken line) and heat pulse velocity (HPV, continuous line) measured in *A. cordata* on July 5 and 6 (a) and the cross-correlation function between their temporal series (b).

The correlation between HPV and UAE was low ($r < 0.5$) and the cross-correlation function demonstrated a phase shift between the two temporal series (Fig. 5b).

In 1990 the measurements on *A. cordata* were performed in the period 8 to 11 October; during these days the air temperature was between 13 and 25 ^{0}C, the shortwave solar radiation up to 500-550 W m^{-2}, and the vapour pressure deficit up to 2.0 kPa.

The xylem water potential and the stomatal conductance were both measured on October 11. The water potential decreased from a dawn value of -0.44 MPa to -1.4 MPa at 13:00 h, recovering slightly in the afternoon (Fig. 6a).

The stomatal conductance increased throughout the morning, up to a peak of 0.58 s cm^{-1} at 13:30 h, and decreased in the afternoon (Fig. 6b).

On 8 October UAE began to increase at sunrise from roots and branches, and about 5 hours later from the stem. On the following days, less evident UAE peaks were recorded only from branches and roots (Figs. 7 b-d). During the night UAE were zero from the stem and less than 0.5 ct min^{-1} from roots and branches.

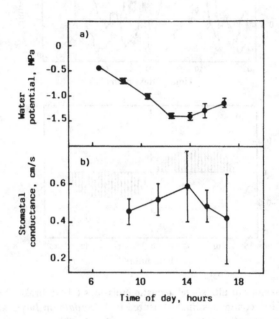

FIG. 6. Daily behaviour of water potential (a) and stomatal conductance (b) measured on October 11 1990 in *A. cordata*.

The maximum correlation coefficients between sap velocity and UAE were 0.4, 0.55 and 0.7, respectively for the stem, the branches and the roots. A time lag of about one hour was estimated between the temporal series of sap velocity and UAE measured on branches and roots (Figs. 8 a-c). However, in the case of the stem a higher correlation ($r = 0.65$) between UAE and sap velocity was computed when the individual values of sap velocities, measured at different depths in the xylem, were considered instead of their average value.

A better understanding of the relationship between water flow and ultrasound emission may be achieved by plotting UAE against sap velocity values. This is done in Figs. 9 a-c.

A rather clear threshold-type relationship is shown: UAE started from the stem when sap velocity was higher than 0.2 m h⁻¹; and from roots and branches when sap velocity was higher than 0.16 m h⁻¹.

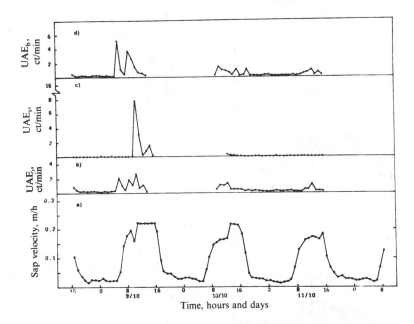

FIG. 7. Daily behaviour of sap velocity (average of four probes) as measured in *A. cordata* in the period 8 to 11 October 1990 (a); daily behaviour of ultrasound acoustic emission (UAE) measured during the same period in *A. cordata* on roots (b), stem (c) and branches (d). In the case of UAE measurements, each point represents the average of two values for the roots and branches, and of four values for the stem.

DISCUSSION

The relationship between acoustic emissions and cavitation of xylem conduits has been discussed by several authors (Sandford & Grace, 1985; Ritman & Milburn, 1988; Tyree & Sperry, 1989a) and there is good agreement that emission and cavitation rates are positively correlated.

In the 1989 experiment, variable acoustic emission rates were recorded from the species that were studied under the same environmental conditions, with higher rates in *Q. pubescens* and lower ones in *Q. ilex*.

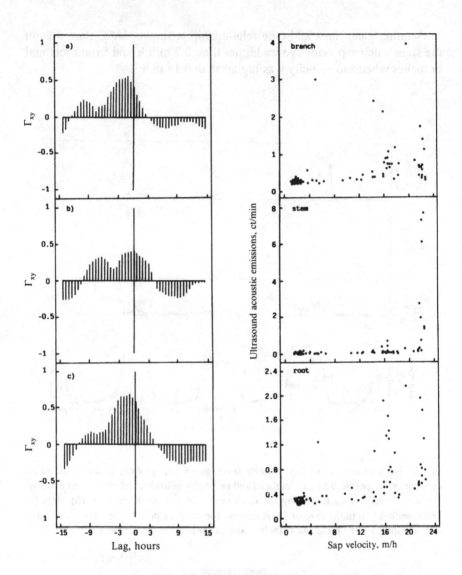

FIG. 8. (left) Cross-correlation function between the temporal series of ultrasound acoustic emission (UAE) and sap velocity (average of four probes) measured in *A. cordata* in the period 8 to 11 October 1990. a) UAE measured on branches; b) on the stem; c) on roots. Negative lag means that UAE happen first.

FIG. 9. (right) Relationships between ultrasound acoustic emission (UAE), as measured on roots (a), stem (b) and branches (c), and sap velocity (average of four probes) as measured on the stem in the period 8 to 11 October 1990 in *A. cordata*.

These differences may indicate higher cavitation rates and vulnerability to embolism in *Q. pubescens*, whose large vessels could enhance the water transport efficiency but decrease the safety of the xylem (Zimmermann, 1983). On the other hand, it is hard to exclude that different emission rates might be caused by different listening distances or sensitivities of the transducer in different wood materials (cf. Sandford & Grace, 1985; Tyree & Sperry, 1989b).

Heat pulse velocities were also variable among the studied species during the 1989 experiment, with the highest velocities recorded still being in *Q. pubescens*; however, when interpreting and comparing absolute values uncertainties may arise even in this case, also because this method is more suitable for diffuse porous species as it assumes that wood is essentially homogeneous.

For these reasons, a comparison between absolute values seems to be an unreliable method for analyzing the results, whereas a discussion on the time patterns and on the relationships between the phenomena seems to be more fruitful.

A good correspondence was found between the patterns of sap velocity and cavitation rate. Overall, this confirms previous experimental evidence, available for both herbaceous and woody plants under field conditions, that cavitation of xylem conduits is associated with high transpiration rates and low water potentials (Tyree *et al.*, 1986; Peña & Grace, 1986; Borghetti, Raschi & Grace, 1989).

Indeed, if water transport in plants is interpreted following Ohm's analogy and Van den Honert's concept (1948), the increase of sap velocity and transpiration rate means that the tension in the xylem has increased. This determines conditions favourable to the cavitation of xylem conduits, in accordance with the air-seeding hypothesis (Sperry and Tyree, 1988; Sperry *et al.*, 1988).

The clear association between the decrease of sap velocity and cavitation rate agrees with the claim by Tyree (1989), that an important role of the stomatal control of transpiration is not only to prevent desiccation damage but also to avoid the so-called "runaway embolism".

Temporal lags between ultrasound emission and sap velocity were observed in some cases: on the morning of July 14 in *Q. ilex,* when HPV rose two hours after UAE; in *A. cordata* on July 5 1989, when the peak of HPV lagged about two hours behind the peak of UAE in the morning and one hour behind in the night; still in *A. cordata* on October 9 1990 when the UAE from the stem lagged 5 hours behind the onset of sap flux. Several factors may account for these lags, including the fact that cavitation of xylem conduits may be a rather patchy phenomenon and that different xylematic volumes might have been sensed by the UAE and HPV transducers.

No variations of flow rates immediately after the ultrasound emission peaks were observed. This may indicate that xylem conductivity was not affected by vessel embolization over short time intervals. Indeed, in a previous experiment appreciable changes of xylem conductivity were evident only after rather long drought periods, during which high cavitation rates were recorded (Borghetti, Raschi & Grace, 1989).

The peaks of both UAE and HPV recorded during the night in *A. cordata* on July 1989 may be caused by nocturnal opening of stomata (cf. Mansfield, 1965), which has been documented as a rather common phenomenon in plants growing in mediterranean environments (Gindel, 1973). It may be also attributed to reallocation of water between parts of the plant experiencing different water potentials. Indeed, the cavitation of xylem conduits in the roots or in the stem could provide sources of water for the rehydration of twigs and leaves (Tyree & Dixon, 1983; Dixon, Grace & Tyree, 1984). No night UAE and sap velocity peaks were recorded in October 1990, when the xylem sources could have been progressively depleted. Indeed, at this date the concurrent rise of sap flow and cavitation in branches indicates a simultaneous onset of crown transpiration and stem flow, which denotes a lack of stem capacitance (cf. Schulze *et al.*, 1985).

The threshold-type relationships observed, in *Alnus cordata*, between cavitation rate and water flow, supposed to be positively correlated with the tension in the xylem, seem to be in accordance with the hypothesis that a critical pressure difference through the intervessel pit membrane is required in order to determine the aspiration of air bubbles within the xylem conduits (Tyree & Dixon, 1986; Sperry & Tyree, 1988; Sperry, Tyree & Donnelly, 1988).

It is evident that long-term experiments are necessary for a deeper understanding of the water transport processes in woody plants. In particular, a useful development of the research could be based on the definition of *in vivo* "vulnerability curves", showing how embolism increases as water potential decreases, by means of concurrent and continuous measurements of xylem water potential, cavitation rates and water flow on different plant parts (roots, stem, major and minor branches). These measurements may provide the means to test the plant segmentation hypothesis (Zimmermann, 1983), validate the models that have been recently proposed (Tyree, 1988), and better evaluate if and how in woody plants the xylematic water transport is vulnerable to the proposed phenomenon of runaway embolism, over daily and seasonal time intervals (Tyree & Sperry, 1988; Tyree, 1989).

ACKNOWLEDGMENTS

This research was supported by the National Research Council of Italy.

REFERENCES

Borghetti, M., Raschi, A. & Grace, J. (1989) Ultrasound emission after cycles of water stress in *Picea abies. Tree Physiology,* **5,** 229-237.

Dixon, M.A., Grace, J. & Tyree, M.T. (1984) Concurrent measurements of stem density, leaf water potential, stomatal conductance and cavitation on a shoot of *Thuja occidentalis* L. *Plant, Cell and Environment,* **7,** 615-618.

Gindel, I. (1973) *A new ecophysiological approach to forest-water relationships in arid climates.* Dr. W. Junk B.V. Pub., Holland.

Hüber, B. & Schmidt, E. (1937) Eine Kompensationsmethode zur thermoelektrishen Messung Langsamer Saftstrome. *Ber. dt. Bot. Ges.,* **55,** 514-529.

Mansfield, T.A. (1965) Studies in stomatal behaviour. XII. Opening in high temperature in darkness. *Journal of Experimental Botany,* **16,** 721-731.

Marshall, D.C. (1958) Measurement of sap flow in conifers by heat transport. *Plant Physiology,* **33,** 385-396.

Milburn, J.A. (1979) *Water flow in plants.* Longman, London.

Milburn, J.A. & Johnson, R.P.C. (1966) The conduction of sap. II. Detection of vibrations produced by sap cavitations in *Ricinus* stem. *Planta,* **69,** 43-52.

Pearcy, R.W., Schulze, E.-D. & Zimmermann, R. (1989) Measurement of transpiration and leaf conductance. In *Plant Physiological Ecology, Field Methods and Instrumentation* (ed. R.W. Pearcy, J. Ehleringer, H.A. Mooney & P.W. Rundel), pp. 137-160. Chapman and Hall, London.

Peña, J. & Grace, J. (1986) Water relations and ultrasound emissions of *Pinus sylvestris* before, during and after a period of water stress. *New Phytologist,* **73,** 861-871.

Ritman, K.T. & Milburn, J.A. (1988) Acoustic emission from plants: ultrasonic and audible compared. *Journal of Experimental Botany,* **39,** 1237-1248.

Sandford, A.P. & Grace, J. (1985) The measurement and interpretation of ultrasounds from woody stems. *Journal of Experimental Botany,* **36,** 298-311.

Schulze, E.-D., Cermák, n, J., Matyssek, R., Penka, M., Zimmermann, R., Vasicek, F., Gries, W. & Kucera, J. (1985) Canopy transpiration and water fluxes in the xylem of the trunk of *Larix* and *Picea* trees - a comparison of xylem flow, porometer and cuvette measurements. *Oecologia,* **66,** 475-483.

Sperry, J.S. & Tyree, M.T. (1988) Mechanism of water stress-induced embolism. *Plant Physiology,* **88,** 581-587.

Sperry, J.S., Tyree, M.T & Donnelly, J.R. (1988) Vulnerability of xylem to embolism in a mangrove vs. an inland species of Rhizophoraceae. *Physiologia Plantarum*, **74**, 276-283.

STSC Company (1985) StatGraphic, Statistical Graphics System. Rockville, Maryland, USA.

Swanson, R.H. & Whitfield, D.W.A. (1981) A numerical analysis of heat pulse velocity: theory and practice. *Journal of Experimental Botany*, **32**, 221-239.

Tyree, M.T. (1988) A dynamic model for water flow in a single tree: evidence that models must account for hydraulic architecture. *Tree Physiology*, **4**, 195-217.

Tyree, M.T. (1989) Cavitation in trees and the hydraulic sufficiency of woody stems. *Annales Sciences Forestiers*, **46 suppl.**, 330s-337s.

Tyree, M.T. & Dixon, M.A. (1983) Cavitation events in *Thuya occidentalis*? Ultrasonic acoustic emissions from the sapwood can be measured. *Plant Physiology*, **72**, 1094-1099.

Tyree, M.T. & Dixon, M.A. (1986) Water stress induced cavitation and embolism in some woody plants. *Physiologia plantarum*, **66**, 397-405.

Tyree, M.T., Fiscus, E.L., Wullschlegel, S.D. & Dixon, M.A. (1986) Detection of xylem cavitation in corn under field conditions. *Plant Physiology*, **82**, 597-599.

Tyree, M.T. & Sperry, J.S. (1988) Do woody plants operate near the point of catastrophic xylem disfunction caused by dynamic water stress. *Plant Physiology*, **88**, 574-580.

Tyree, M.T. & Sperry, J.S. (1989a) Vulnerability of xylem to cavitation and embolism. *Annual Review of Plant Physiology and Molecular Biology*, **40**, 19-38.

Tyree, M.T. & Sperry, J.S. (1989b) Characterization and propagation of acoustic emission signals in woody plants: towards an improved acoustic emission counter. *Plant, Cell and Environment*, **12**, 371-382.

Van den Honert, T.H. (1948) Water transport in plants as a catenary process. *Discussion Faraday Society*, **3**, 146-153.

Zimmermann, M.H. (1983) *Xylem Structure and the Ascent of Sap*. Springer Verlag, Berlin.

NMR and water transport in plants

S. RATKOVIC[1] AND G. BACIC [2]

[1] Dept. of Technology and Chem. Res., Maize Research
Institute, 11080 Zemun-Beograd, and [2] Dept. of Phys.
Chem., Faculty of Science, University of Beograd,
Beograd, Yugoslavia.

SUMMARY

NMR methods have a unique capability for the studying of water relations in biological systems by virtue of their being noninvasive and nondestructive.

We will show here the potential use of the proton NMR technique for studying all aspects of water relations of plants including: (1) water exchange across the membranes of both single cells (algae) and cells within tissues, (2) radial water transport (short distance) and controlling steps of water transport in root tissue, and (3) xylem (long distance) transport of water in plants.

In addition we will demonstrate the potential of magnetic resonance imaging (MRI) in studying different aspects of water relations of plants by observing water as seen in a MR image of a cross section through plant tissue (stem or root) obtained noninvasively.

INTRODUCTION

The basic discoveries of nuclear induction made in 1946 laid down the fundamentals of the nuclear magnetic resonance (NMR) technique, which has had an explosive development, with application in all branches of physics, chemistry and biology. The sophisticated ^{13}C and ^{31}P NMR spectroscopy has been successfully applied to studies of metabolism and transport processes in plants (Loughman & Ratcliffe, 1984; Roberts, 1987).

Since our interest here is in water transport in plants we shall consider those NMR techniques which can measure resonance and relaxation of hydrogen nuclei, *i.e.* protons, in water molecules, and will also show that the proton NMR signal of water in plants is dependent on the state of water in the tissue and on the motion of the fluid within the NMR probe.

FUNDAMENTALS OF NUCLEAR MAGNETIC RESONANCE

The basic principles of NMR and the associated relaxation phenomena can be explained using a simple vector model shown in Fig.1 (Farrar & Becker, 1971). If a sample containing hydrogens, *i.e.* water molecules (plant stem or root, for instance) is put in a constant magnetic field B_o (Fig. 2) a macroscopic magnetization M is produced as a result of superposition of all proton magnetic moments μ_i per unit volume ($M = \Sigma \mu_i$). Under the action of B_o, the magnetization M will precess around the z-axis (B_o direction) with a characteristic frequency, the so-called Larmour frequency $\nu_L = (\gamma/2\pi) \cdot B_o$, which is proportional to the magnetic field B_o (γ is a gyromagnetic constant for protons). The projection of the magnetization along z-axis is called M_z and along the other two axes M_{xy} (Fig. 1). An alternating field with frequency ν_o and strength $B_1 \ll B_o$ can be applied along the x-axis by using a radiofrequency coil (Figs. 1 & 2), and when $\nu_o = \nu_L$ the resonance condition is fulfilled and the system of magnetic moments will absorb energy from the magnetic field. What actually happens is that the vector M_z will be tipped away from the z-direction for a certain angle depending on the duration of the B_1 field action. If it acts in short pulses M_z can be nutated to the *xy*-plane ($\pi/2$ or 90° pulse) or to the -z direction (π or 180° pulse). After cessation of the pulse the whole spin system is not in equilibrium and the components M_z and M_{xy} return back to their equilibrium values $M_z = M_o$ and $M_{xy} = 0$ during time t. This is described by two relaxation processes with time constants T_1 and T_2, called spin-lattice and spin-spin relaxation time, respectively:

$$M_z = M_0[1\text{-}\exp(\text{-}t/T_1)] \qquad \text{Equation 1}$$

$$M_{xy} = M_0\exp(\text{-}t/T_2) \qquad \text{Equation 2}$$

T_1 relaxation is due to an energy exchange between the spin system and surroundings, while the T_2 process describes interaction within the spin system itself (spin exchange). Both parameters are highly dependent on molecular motion as affected by temperature, flow, etc.

The signal induced in the radiofrequency coil after cessation of the pulse is the free induction decay signal (FID).

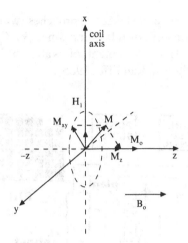

FIG.1. Vector model of the NMR experiment.

After a Fourier transformation such a signal can be transformed from the time domain to the frequency domain, *i.e.* to obtain a NMR spectrum. Measurement in the time domain results in the parameters T_1 and T_2 directly, while in the frequency domain the line width can be indirectly correlated with T_2 ($\Delta v = 1/\pi T_2$), while the position of the line reports on the immediate proton environments (*e.g.* H_2O vs CH_2 protons). According to this, liquid water gives a sharp proton NMR line (long T_2), while the line of water sorbed on a surface or containing paramagnetic impurities is broad (short T_2).

The contemporary magnets used in high resolution NMR spectroscopy exploit the effect of superconductivity at liquid He temperature. They produce a stable highly homogenous field up to 14 tesla, corresponding to a proton resonance of 600 MHz. Although the sensitivity increases with the strength of the field, these magnets are not suitable for *in vivo* experiments because of their narrow bore. The NMR relaxometers used in experiments described here have classical electromagnets with an iron core (Fig. 2).

SHORT DISTANCE TRANSPORT OF WATER

The processes of water diffusion at the cellular level including water exchange across membrane as well as water transport in tissue (radial water movement in roots) can be effectively studied by NMR techniques (Bacic, Srejic & Ratkovic, 1990). The main problem is to distinguish proton signals from different compartments - extracellular and intracellular water. Several NMR approaches were applied to these problems: (1) measurement of relaxation times (T_1, T_2), (2) chemical shift measurement, and (3) measurement of water diffusion coefficient D (Ratkovic, 1981; Belton & Ratcliffe, 1985).

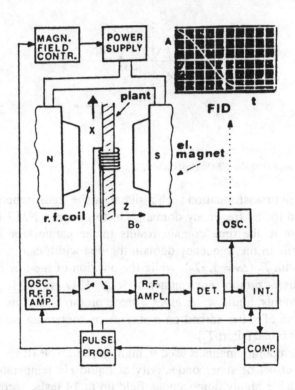

FIG. 2. Typical block scheme of an NMR relaxometer with an electromagnet and a radiofrequency coil. The shape of a FID signal of water in tissue as seen on the oscilloscope screen is shown in the right upper corner. Experiments with long distance transport or MRI (Magnetic Resonance Imaging) require additional magnetic field gradient coils. A segment of a plant stem is shown within the radiofrequency coil.

Water exchange time (τ_i) across the membrane has been measured in many single cell systems or in tissues using the modified NMR method of Conlon & Outhred (1972) with paramagnetic doping (Mn^{2+}) of the extracellular space. The relaxation time of the intracellular water T_1 becomes shorter (T_1') after addition of paramagnetic centers outside the cells (impermeable ions or molecules: MnEDTA, GdDTPA) and after certain approximations the water exchange time τ_i can be calculated from:

$$\tau_i = (T_1 \, T_1') \, / \, (T_1 - T_1')$$ Equation 3

and the diffusional permeability of a cell P_d is:

$$P_d = V \, / \, (A\tau_i)$$ Equation 4

where V and A are volume and surface area of the cell. The NMR method has a distinct advantage over the other methods because it definitely eliminates the problem of an unstirred layer near membranes encountered with these methods.

Using T_1 measurements in the presence of $MnCl_2$ we obtained $P_d = 2.5 \; 10^{-3}$ cm s^{-1} for *Nitella mucronata* cells (Ratkovic & Bacic, 1980) and $P_d = 6.5 \; 10^{-3}$ cm s^{-1} for maize cortical cells (Bacic & Ratkovic, 1984) in agreement with results for other plant systems (Bacic, Srejic & Ratkovic, 1990).

Radial exchange of water in tissue can be studied using H_2O-D_2O exchange and measuring build-up of the proton NMR signal in D_2O solution surrounding the tissue (Fig. 3). We have analysed such exchange in maize roots (Bacic & Ratkovic, 1987) using three different models: (1) two compartments in series, (2) diffusion within apoplast with simultaneous permeation through the cell membrane, and (3) cylinder diffusion.

This approach resolved the existing controversy about the main pathways for radial flow of water (Newman, 1976) showing that the major radial water flux in the root occurs through the cells.

LONG DISTANCE TRANSPORT OF WATER

Special NMR repetitive pulse techniques which use the effect of water diffusion along magnetic field gradients were developed for measurement of water flow along xylem vessels of certain plants (van As, 1982; van As & Shaafsma, 1984; Reinders, 1987). During the measuring procedure the main stationary water phase is not detected. This can be done practically by introducing gradient coils along the x-axis for instance, in an arrangement like that in Fig. 2. Typical NMR response curves obtained with the same plant and for three different flow rates are shown in Fig. 4 where the curve A corresponds to fastest flow and the curve C to the slowest flow. Two characteristic parameters of such a curve are: t_m (time corresponding to S_{max}) and $S_0 = [dS(t)/dt]_{t=0}$ (initial slope of the curve at $t=0$).

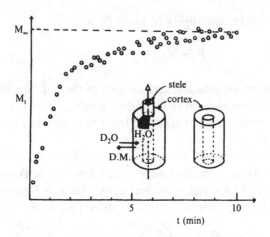

FIG. 3. Build-up of the "proton" FID signal of H_2O in the D_2O solution bathing root segments of *Zea mays* L. Dextran-magnetite was present in the external solution.

These parameters can be related to linear flow rate v (m s⁻¹) and volumetric flow Q (m³ s⁻¹):

$$V = K_v/t_m \qquad \text{Equation 5}$$

$$Q = K_q S_0 = K_q [dS(t)/dt]_{t=0} \qquad \text{Equation 6}$$

In these equations K_v and K_q are calibration constants which depend on the following parameters: magnetic field gradient (G), time interval between two radiofrequency pulses (τ), geometry of the radiofrequency coil, flow profile and T_1 and T_2. The flow rates Q and v are related via the effective cross sectional area for flow A:

$$Q = v A \qquad \text{Equation 7}$$

It was shown that the RP NMR method was not sensitive for plants with narrow xylem vessels (diameter of xylem vessels 90 μm) having short T_2, but was successfully applied to plants like cucumber, pumpkin and gherkin (Reinders, 1987).

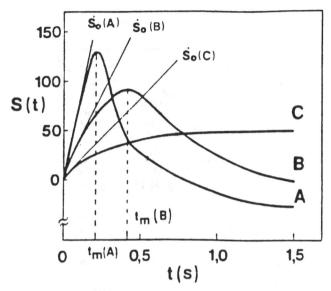

FIG. 4. Shape of the NMR response curves for three different flows of water along xylem with characteristic parameters t_m and S_0 used to calculate v and Q.

MAGNETIC RESONANCE IMAGING OF WATER IN PLANTS

The water content of a plant tissue depends on its morphology and physiological status. It is a general conclusion from the proton NMR relaxation measurement of tissues that T_1 and T_2 are in positive correlation with water content (this can be changed by the presence of paramagnetic centers). Heterogeneity of these parameters in tissue is shown in Figs. 5a & 5b for different parts of *Zea mays* L. roots - tip, stele, cortex, whole root (Bacic & Ratkovic, 1987).

This is a rough picture but we can say that each point within the tissue can be characterized by its water content, T_1 and T_2 (Fig. 6).

The images of heterogeneous systems like plants can be obtained using the MRI technique, which is based on the application of linear magnetic field gradients along different directions across a sample (Hutchison, 1984; Morris, 1986). Such a gradient (G_x) can be produced by additional coils in an arrangement like that in Fig. 2, and if it acts along the x-direction the main magnetic field along the z-direction is:

$$B_z = B_0 + xG_x \qquad \text{Equation 8}$$

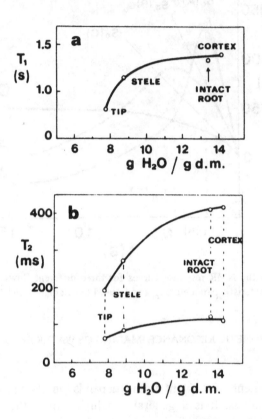

FIG.5. Spin relaxation times of water in different parts of maize roots: (a) T_1 vs amount of water per gram dry matter, and (b) T_2 (two components) vs amount of water per gram dry matter.

and the Larmour frequency which is normally $v_L = (\gamma/2\pi)B_0$ is:

$$v_L = (\gamma/2\pi)B_z = (\gamma/2\pi)B_0 + (\gamma/2\pi)B_x \qquad \text{Equation 9}$$

After application of a 90° pulse to the magnetization the resulting FID signal will not depend on a single frequency but rather on a whole spectrum of frequencies, and each component will correspond to a particular nuclear species in the corresponding magnetic field. A strong magnetic field gradient applied along the x-direction (cf. Fig. 2) is used to select the imaging plane (slice) perpendicular to the gradient direction. This is analogous to cutting a cross-sectional area of a plant stem at certain levels. Two additional gradients (perpendicular to the x-gradient) are used to obtain the spatial distribution of water within the selected slice. The intensity of a particular structure seen in the MR image depends on the intrinsic tissue parameters such as its T_1, T_2 and water content (proton density) as well as on the imaging sequence selected (Fig. 6). There are different imaging methods in use now but they are all based on the same general principle.

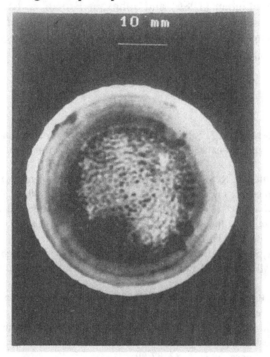

FIG. 6. A transverse proton density image of the *Dracaena* stem (slice thickness 6 mm, refocusing time TE = 15 ms, repetition time TR = 1.6 s). The image was obtained at 63 MHz on Siemens Magnetom medical imager (MR Center, University Clinical Center, Belgrade) using a standard orbit coil.

The efforts of scientists were directed both to get images of large objects with relatively low spatial resolution and to get images of small objects with high spatial resolution (micro imaging or NMR microscopy). It was possible to get an image of a single cell with an in-plane pixel (unit element in two dimensions) resolution of 10 x 13 μm (250 μm slice thickness) on a 400 MHz NMR imager (Aguayo *et al.,* 1986). There have been several interesting MRI studies on plant systems. Bottomley, Rogers & Foster (1986) and Brown, Johnson & Kramer

(1986) described imaging of plant roots at a spatial resolution of 100 μm and showed dependence of the pixel signal intensity on the water content during transpiration. More detailed study of the roots was done by Connelly et al. (1987) using mini-imaging at 200 MHz. These authors demonstrated the potential of the MRI to study water exchange (D_2O) or uptake of paramagnetic ions (Mn^{2+}) by plant roots - in agreement with our previous proton NMR relaxation experiments.

Eccles & Callaghan (1986) and Johnson, Brown & Kramer (1987) used MR microscopy to study plant stem, while Hall et al. (1986) and Olson, Chang & Wang (1990) demonstrated the application of MRI to morphological studies on large samples of woody plants.

REFERENCES

Aguayo, J. B., Blackband, S. J., Schoeniger, J., Mattingly, M. A. & Hintermann, M. (1986) Nuclear magnetic resonance imaging of a single cell, *Nature* (London), **322**, 190-191.

Bacic, G. & Ratkovic, S. (1984) Water exchange in plant tissue studied by proton NMR in the presence of paramagnetic centers. *Biophysical Journal*, **45**, 767-776.

Bacic, G & Ratkovic, S. (1987) NMR studies of radial exchange and distribution of water in maize roots: the relevance of modelling of exchange kinetics. *Journal of Experimental Botany*, **38**, 1284-1297.

Bacic,G., Srejic, R. & Ratkovic, S. (1990) Water transport through membranes. A review of NMR studies of model and biological membranes. *Studia biophysica*, **138**, 95-104.

Belton, P.S. & Ratcliffe, R.G. (1985) NMR and Compartmentation in Biological Tissues. *Progress in NMR Spectroscopy,* **17**, 241-279.

Bottomley, P. A., Rogers, H. H. & Foster, T. H. (1986) NMR imaging shows water distribution and transport in plant root systems *in situ*. *Proceedings of the National Academy of Sciences of USA*, **83**, 87-89.

Brown, J. M., Johnson, G. A. & Kramer, P. J. (1986) *In vivo* Magnetic Resonance Microscopy of changing water content in *Pelargonium hortorium*. *Plant Physiology*, **82**, 1158-1160.

Conlon, T. & Outhred, R. (1972) Water diffusion permeability of erythrocytes using a nuclear magnetic resonance technique. *Biochimica et Biophysica Acta*, **288**, 354-361.

Connelly, A., Lohman, J.A.B., Loughman, B.C., Quiquampoix, H.D. & Ratcliffe, R. G. (1987) High resolution imaging of plant tisssue by NMR. *Journal of Experimental Botany*, **38**, 1713-1723.

Eccles, C.D. & Callaghan, P.T. (1986) High resolution imaging. The NMR microscope. *Journal of Magnetic Resonance*, **68**, 393-398.

Farrar, T.C. & Becker, E.D. (1971) *Pulse and Fourier Transform NMR*. Academic Press, New York.

Hall, L.D., Rajanayagam, V., Stewart, W.A., Steiner, P.R. & Chow, S. (1986) Detection of hidden morphology of wood by magnetic resonance imaging. *Canadian Journal of Forest Research*, **16**, 684-687.

Hutchison, J.M.S. (1984) NMR proton imaging techniques. In *Magnetic Resonance in Medicine and Biology* (ed. M.A. Foster), pp 173-190, Pergamon Press, Oxford.

Johnson, G.A., Brown, J. & Kramer, P.J. (1987) Magnetic resonance microscopy of changes in water content in stems of transpiring plants. *Proceedings of the National Academy of Sciences of USA*, **84**, 2752-2735.

Loughman, B.C. & Ratcliffe, R.G. (1984) Nuclear magnetic resonance and the study of plants. In *Advances in Plant Nutrition* (ed. P.B. Tinker & A. Lauchli), Vol.1, pp. 240-283, Praeger, New York.

Morris, P.G. (1986) *Nuclear Magnetic Resonance Imaging in Medicine and Biology*. Clarendon Press, Oxford.

Newman, E.I. (1976) Water movement through root systems. *Philosophical Transactions of Royal Society London*, **Ser.B**, **273**, 463-478.

Olson, J. R., Chang, S. J. & Wang, P. C. (1990) Nuclear magnetic resonance imaging: a noninvasive analysis of moisture distributions in white oak lumber. *Canadian Journal of Forest Research*, **20**, 586-591.

Ratkovic, S. (1981) NMR studies of water in biological systems at different levels of their organization. *Scientia Yugoslavica*, **7**, 19-54.

Ratkovic, S. & Bacic, G. (1980) Water exchange in *Nitella* cells: A PMR study in the presence of paramagnetic Mn^{2+} ions. *Bioelectrochemistry and Bioenergetics*, **7**, 405-412.

Reinders, J.E.A. (1987) A nuclear magnetic resonance study of plant-water relationships. PhD thesis, Agricultural University, Wageningen.

Roberts, J.K.M. (1987) NMR in plant biochemistry. In *The Biochemistry of Plants* (ed. P.K. Stumpf & E.E. Conn), Vol. 13, pp. 181-227, Academic Press, New York.

van As, H. (1982) NMR, water and plants. PhD thesis, Agricultural University, Wageningen.

van As, H. & Schaafsma, T.J. (1984) Noninvasive measurement of plant water flow by nuclear magnetic resonance. *Biophysical Journal*, **45**, 469-473.

The symplast radial-axial water transport in plants: a NMR approach

A.V. ANISIMOV

Kazan Institute of Biology, 420111, P.O.B.30, Kazan, Russia.

SUMMARY

We explore the possibility of intercellular water transport through the symplast system by measuring the effective water diffusion coefficient by pulsed NMR after preliminary doping of the intercellular space by paramagnetic ions. Part of the transpirational water flow is shown to occur through the plant symplast. The relation between *quasi* diffusion coefficients and flow velocities at laminar and piston flows is considered for the model of the linear cell chain.

INTRODUCTION

A number of papers have been published concerning the theoretical estimates of the efficiency of symplast as the water pathway in plants. The authors assume that the symplast is the most important pathway for the intercellular transport of water and other substances (Arisz, 1956; Tyree, 1970; Newman, 1976). This assumption permits us to account for the dependence of tissue water regime on inhibitors and activators of metabolism better than other explanations. Hitherto, there have been no direct measurements of symplastic water transport due to the absence of any method that differentiates symplast transfer against a background of apoplast and transmembrane transport. The development of nuclear magnetic resonance (NMR) technique allowed us for the first time to distinguish the symplast component from the total intercellular transfer (Anisimov *et al.*, 1983).

The principle of measuring water diffusion just through the symplast is based on the relaxation suppression of the apoplast water NMR signal by doping the intercellular space with paramagnetic ions of high relaxation efficiency, such as manganese and gadolinium. In other words the water molecules of the intercellular space become invisible, due to the fast relaxation on the paramagnetic ions, while the water transfer from cell to cell through the plasmodesmata remains observable, due to the absence of the direct contacts between the symplast water molecules and paramagnetic ions.

Several papers are devoted to the application of NMR to the diffusion measurements (Torrey, 1953; Stejskal & Tanner, 1965; Tanner, 1970; Cooper *et al.*, 1974). In brief, the pulsed NMR sensitivity to the translational motion of water molecules is connected with the frequency dependence of the Larmor precession of nuclei (protons) on the magnitude of the applied magnetic field. If we "label" the studied sample by a magnetic field gradient, then the spatial displacement of water molecules due to the diffusion or flow in the sample results in change of the precession frequency and loss of the spin packet phase coherence, and these finally result in change of the amplitude of the magnetization signal of H_2O. As a result, the pulsed NMR allows us to measure the intracellular self diffusion coefficient of water, the diffusion coefficient of water in the tissue (D_{ef}), the transmembrane exchange rate, the velocity of xylem water flow, the water diffusion coefficient in the heterogeneous system when current is applied (*quasi* diffusion coefficient, D') and, what is important, distinguishes diffusion from current (Anisimov, 1982; van As & Scha'afsma, 1984).

MATERIALS AND METHODS

The pulsed NMR method *spin-echo* with the pulsed magnetic field gradient was used to study the symplast water transfer in experiments described below (Hahn, 1950; Tanner, 1970).

The measurements during the study of symplast water radial transport were carried out on the segments of the suction zone of 7-day-old maize roots (var. "Sterling"). A magnetic field gradient was directed across the segments. The axial transfer was studied on the roots of 20-day-old maize shoots. The magnetic field gradient was directed along the roots. The measured zone 5 mm long was in the center of the suberized part of the root, the full length of which was 15-20 cm.

RESULTS AND DISCUSSION

The measurement of the radial water diffusion after the preliminary doping of the intercellular space of maize roots with paramagnetic ions, manganese or gadolinium, allowed us to ascertain the high degree of connectedness of the cells through the plasmodesmata: 60-80% of water can diffuse from cell to cell through plasmodesmata (Anisimov *et al.*, 1983). The plasmodesmata permeability is illustrated in Fig. 1.

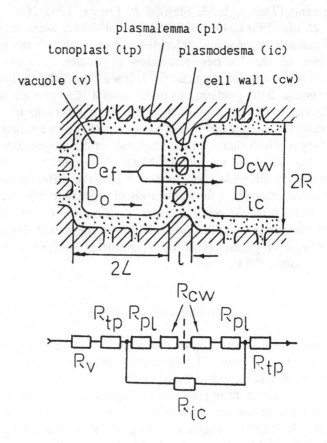

FIG.1. The equivalent scheme of the resistance (R*) to transfer between adjacent cells. See text for further explanations.

The relation between D_{ef} and the parameters of the transfer chain is given by:

$$D_{ic} = \{1[(2L + 1/D_{ef}) - 2(1/P_{tp} + 1/P_{pl} + 1/D_{cw})]\}/$$
$$[2(1/P_{pl} + 1/D_{cw})(2L/D_0 - 2/P_{tp} - (2L + 1)/D_{ef})]$$

Equation 1

where P_{pl} and P_{tp} are water permeability coefficients of the plasmodesmata and tonoplast, respectively, D_{ic} and D_{cw} are effective water diffusion coefficients through plasmodesmata and cell wall respectively, D_0 is the water self diffusion molecular coefficient, L is cell length.

In the case of maize roots the value of D_{ic} for the present scheme is 0.5×10^{-7} cm^2 s^{-1} ($D_{ef} = 0.2 \times 10^{-5}$, $P_{pl} = P_{tp} = 10^{-2}$ cm s^{-1} and the ratio of relative flows through plasmodesmata and plasmalemma is 4:1). The relative plasmodesmata density per cell is 0.01, formally corresponding to a permeability of the order 10^{-2} cm s^{-1}. The latter value is comparable to or somewhat higher than the value for the plasmalemma permeability, which is in the range of 10^{-2} - 10^{-3} cm s^{-1} for various plants (Stout, Cotts & Steponkus, 1977; Bacic & Ratkovic, 1984).

The statement of the problem of the possibility of distant axial water transport through the symplast system is justified taking into account the data on the variations of stem thickness, the decrease of tissue water content in the acropetal direction, the additional effect of root pressure and the effect of the removal of gas embolism. These facts can be explained taking into account the assumption that the fraction of the axial transpiration water flow occurs apart from the xylem vessels through the symplast system of the cells surrounding the vessels. In this case the water pathway may be adapted to transform the cell metabolism energy into the kinetic energy of water movement. The measurement of the parameters of flow through the symplast is based on the fact that the small water volume, entering the cell, disturbs the water diffusion of the cell and of the whole symplast system. The disturbed diffusion is detected by the NMR method as a *quasi*-self-diffusion coefficient, D'. The stochasticity of water flow necessary for the diffusion measurement is imposed by the tissue symplast system heterogeneity.

The existence of the axial component was shown by the D' measurement in the suberized part of the root of the 20-day-old *Zea mays* intact shoots (Fig. 2) (Anisimov & Galyaltdinov, 1988).

The modulation of the flow by cutting off the upper driving force (leaves) resulted in the 1.2-1.5 times decrease of D' compared to the value in the norm, the cutting off the lower driving force (root suction zone) results in the 2 times increase of D'. The latter is explained by the removal of the limiting resistance localized in the root suction zone. The hydrodynamic model of water flow through the cell chain is considered to define the relation between D' and the average linear velocity V_0 with the periodical alternation of vacuole zone, membrane zone (tonoplast) and plasmodesmata zone, (Anisimov & Egorov, 1990). The plasmodesmata are presented as hollow capillaries for simplification of the problem and the apoplast water transfer was excluded from consideration. The latter condition can be satisfied in the experiment by doping paramagnetic ions into the tissue intercellular space.

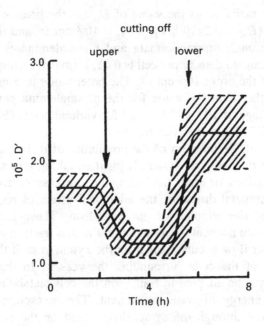

FIG. 2. The dynamics of D' changes in *Zea mays* L. at the successive removal of the upper and the lower water driving forces (dashed region shows the variation of D' values).

The explicit relation between D' and V_0 for the piston flow is given by:

$$D'/D_0 = 1 - 1/2P_e[(\exp P_e - 1)(\exp \delta P_e - 1)]/(\exp P_e(2 + \delta) - 1)$$

$$\text{Equation 2}$$

where $P_e = V_0 L/D_0$, $\delta = 1/mL$, m - the density of plasmodesmata distribution, D_0 - water self diffusion molecular coefficient, l - plasmodesmata length. The more realistic case of Stokes flow admits only numerical solution. The observed dependances of D' on P_e at different α are shown on Fig. 3. It follows from Equation 2 that the D' value can't exceed the D_0 value for the case of piston flow, while in the case of laminar flow the D' value increases with the increase of V_0 (Fig. 3). In the case of maize roots the norm of D' is 1.7×10^{-5} cm^2 s^{-1}, at $m = 0.01$, $L = 40 \times 10^{-4}$ cm, $l = 0.5 \times 10^{-4}$ cm, $D_0 = 2.5 \times 10^{-5}$ cm s^{-1}, the V_0 value for the piston flow is defined as 7×10^{-3} cm s^{-1}, and that for the laminar Stokes flow as 6×10^{-3} cm s^{-1}. Unfortunately, literature data are not available to compare the observed values with the xylem transfer velocity for maize. If we assume the mean velocity of the xylem water transfer is 1 cm s^{-1} for different plants (Slavik, 1974; van As & Scha'afsma, 1984), then the velocity of axial symplast water flow is approximately two orders of magnitude lower than the velocity of the xylem transfer.

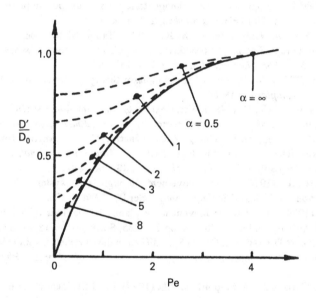

FIG. 3. The dependence of D'/D_0 on $P_e = V_0 R/D_0$ for the laminar water flow at different values of α. $P_e = V_0 R/D_0$, $\alpha = 1 / mR$, R - cell radius, $R / L = 1$.

The final scheme of the distant water transfer assumes radial branching of the water flow into the xylem from the axial symplast flow. This scheme explains the mechanism of the removal of gas embolism in the xylem: the radial branching of water from the symplast flow dissolves or supplants air from the vessels.

ACKNOWLEDGEMENTS

The author would like to thank Dr M. Galyaltdinov for help in measurements and Dr N. Dautova for assistance in preparation of the manuscript.

REFERENCES

Anisimov, A.V. (1982) Pulsed NMR in diffusional water transfer study in biological structures. *Studia Biophysica,* **91,** 1-8.
Anisimov, A.V. & Egorov, A.C. (1990) Otsenka skorosti aksialnogo toka vody po simplastu linejnoj tsepi kletok rastenija. *Biofizika.*
Anisimov, A.V., Evarestov, A.S., Samuilova, I.F. & Gusev, N.A. (1983) Impulsnyj metod JaMR v otsenke megkletochnogo transporta vody po simplastu. *Dokladi Akademii Nauk SSSR Leningrad/Moskva,* **271,** 1246-1249.
Anisimov, A.V. & Galyaltdinov, M.K. (1988) Tangentsialnyj perenos vody o simplastu kornja *Zea mays.* Issledovanie impulsnym metodom JaMR. *Dokladi Akademii Nauk SSSR Leningrad/Moskva,* **289,** 757-760.
Arisz, W.H. (1956) Significance of the symplasm theory for transport in the root of *Zea mays. Protoplasma,* **46,** N1.
Bacic, G. & Ratkovic, S. (1984) Water exchange in plant tissue studied by proton NMR in presence of paramagnetic centers. *Biophysical Journal,* **45,** 767 -776.
Cooper, R.L..,Chang, D.B., Young, A.C., Martin, C.J. & Ancker-Johnson, B. (1974) Restricted diffusion in biological systems. *Biophysical Journal,* **14,** 161-177.
Hanh, E.L. (1950) Spin-echoes. *Physical Review,* **80,** 580-584.
Newman, B.E.I. (1976) Water movement through root system. *Philosophical Transactions of the Royal Society,* London, Ser.B, **273,** 463-478.
Slavik, B. (1974) Liquid water movement in plants. In *Methods of Studying Plant Water Relations.* Ecological Studies, pp 219-235, Springer-Verlag, Berlin.
Stejskal, E.O. & Tanner, J.E. (1965) Spin-diffusion measurements: spin-echoes in the presence of a time-dependent field gradient. *Journal of Chemical Physics,* **342,** 288-292.
Stout,D.G., Cotts, R.M. & Steponkus, P.L. (1977) The diffusional water permeability of *Elodea* leaf cells as measured by nuclear magnetic resonance. *Canadian Journal of Botany,* **55,** 1623-1631.
Tanner, J.E. (1970) Use of the stimulated echo in NMR diffusion studies. *Journal of Chemical Physics,* **52,** 2523-2526.
Torrey, H.C. (1953) Nuclear spin relaxation by translational diffusion. *Physical Review,* **92,** 962-969.
Tyree, M.T. (1970) The symplast concept. A general theory of symplastic transport according to thermodynamics of irreversible processes. *Journal of Theoretical Biology,* **26,** 181-214.
van As, H. & Scha'afsma, T.J. (1984) Noninvasive measurement of plant water flow by nuclear magnetic resonance. *Biophysical Journal,*45, 469-473.

Reproductive adaptation by polyembryony of coniferous forest trees under climatic stress as revealed by the metabolism of tritiated water

D. J. DURZAN

Department of Environmental Horticulture, University of California, Davis, CA 95616-8587, USA.

SUMMARY

Cleavage and budding polyembryony occurs predominantly in conifers adapted to severe environments as found near the Arctic. Cleavage and budding polyembryony are processes that reconstitute multiple proembryos originating by the division or cloning of a single proembryo into a group of identical proembryos. This process provides little selective genetic advantage since all embryos have the same genotype, but it significantly increases the ability for that genotype to survive and to adapt to climatic vagaries by producing multiple viable embryos. With several Norway spruce and pine genotypes from northern latitudes and/or high elevations, the processes of cleavage and budding can be repeated and controlled in bioreactors that serve as an artificial ovule. In the mother tree, nutrients and water are provided to polyembryogenic masses in ovules through the xylem and phloem connections between the ovuliferous scale and the tree. In bioreactors, process controls now consider the effects of climate on the composition of seeds at the seed source. For each seed source, the composition of nutrients derived from the mother tree is reformulated in the culture medium to precondition the polyembryonic process and to contribute to the recovery of viable embryos. Before a zygotic model reference can be developed for the reconstitution process, factors associated with the seed habit and rest period, based on climatic stress, need to be sorted out and identified. Studies with dry seeds of *Pinus banksiana* Lamb. have revealed that covalently labeled tritium, derived from tritiated water, can be recovered from metabolites in the soluble and insoluble phases of tissues during imbibition. The fate of tritiated water is considered in terms of the cell cycle (DNA replication via ribonucleotide reductase) and isotopic effects that inhibit the metabolic processes being studied. This information provides new background concepts and leads to the reconsideration of how conifers adapt to climatic stress through polyembryony and the metabolism of water.

INTRODUCTION

This report describes relationships between extreme climatic variables and the developmental opportunism evident in conifer embryonal-suspensor masses (ESMs) undergoing cleavage and budding as part of the reproductive process. Interannual and seasonal, global climatic changes can now be visualized by earth imagery from satellites (Dept. of Energy, 1989; Houghton, 1984). The changes that are evident between winter and summer in temperate boreal forests dramatically reveal how coniferous forest trees must spend much of their time at rest or dormancy and under considerable temperature and moisture stress.

With the global concern over climatic changes, threats to water quality and availability, threats to biodiversity, increases in the numbers of endangered species, the growing human population with their economic needs, and trends based on new energy and biomass sources, there is a need to maintain adequate stocks and supplies of elite, commercially important conifers.

Where climate is variable and severe, water is often a critical factor associated with tree survival. The susceptibility of climate to change is most often evident at the margins of climatic zones. I will focus mainly on temperate zone, boreal conifers (*Pinus banksiana* Lamb., *Picea glauca* (Moench) Voss. and *Pseudotsuga menziesii* (Mirb.) Franco) that span most of North America, from the east coast to the Rocky Mountains, and from the northern tree line to the southernmost limit of the range in the mid-west and Pacific northwest in the USA. This presentation will also relate to various genotypes of Norway spruce (*Picea abies* Karst.) in Finland. Air pollution has recently contributed to the decline of Norway spruce forests (Schulze, 1989).

The phenotypic variances of populations of conifers and, more specifically, of clones used in tree improvement need to be considered in terms of their xerophytic habit, reproductive biology and potential for mass propagation. Needles of conifers are efficient cloud collectors, which helps them to collect and retain water during the onset of drought conditions. Where water is concerned, poor drainage at the site and low relative humidities are known to affect the hydrogen isotope ratios in cellulose (Yapp & Epstein, 1982). Conifers present a xerophytic habit that is often characterized by narrow growth rings, specialized leaf morphology and precocious reproduction having opportunistic developmental phases.

The mass replanting of trees for reforestation and afforestation and restoration of environmental quality is now widely proposed (*e.g.* FAO, 1985). Unfortunately, for the important soft wood pulp and paper species in all climates, vegetative propagation methods are not always feasible because of the recalcitrance of conifers to respond to this technology. Moreover, seed sources are in short supply and highly variable in terms of climatic influences on flowering.

In a study of conifer seeds, collected from severe environments near the Arctic in Europe and North America, Simak (1973) and Sziklai (1986) have observed that polyembryony is more common compared with those of southern populations. Here, we are not referring to all types of polyembryony (Table 1), but mainly to polyembryony of the cleavage and budding type.

TABLE 1. Products produced by polyembryogenesis as a form of agamospermy (modified from Durzan, 1990a).

Type	Product attributes
Simple	Several zygotes, but one embryo from one zygote, producing groups of embryos of fraternal type (not clones).
Cleavage and budding	Asexual reconstitution of transplanted zygotic embryos (embryonal-suspensor masses) yielding an identical type (clones). One of more groups of a given fraternal type are possible. Cleavage and budding are similar except in terms of source and topological orientation of the reconstituted embryos.
Sporophytic	Adventitious embryos arise by budding from the nucellus (clones of the mother tree) or from differentiated embryonal cells (clones of the new generation).
Gametophytic	The haploid embryos arise from transformed female gametophytes, *i.e.*, they are products of meiosis and are not considered clones. If embryos become diploid, they are not identical to the parental tree.

In describing relationships between extreme climatic variables related to water and the reproductive processes of conifers, I will emphasize:
1) genetics, and especially population genetics, with reference to climates at seed sources;
2) the new generation in the developing seed, as revealed by cell and tissue culture studies. Here, the nutrition for the embryo is heterotrophic rather than directly photoautotrophic, as in the mother tree. In both contexts, we need to recognize that the water and nutritional status, established by the mother tree, directly affects the survival and

adaptability of the new generation;

3) the major factor limiting the filling of seed, *viz.* the availability of nitrogen. Boreal forest soils are naturally low in nitrogen. Two to five years may be needed for the tree to build up nutrient reserves to produce viable seeds, even though flowering and cone development occur more frequently (Silen, 1982). The supply of nitrogen and other essential substrates enters the developing seed through vascular connections with the mother tree. In conifers, the mother tree provides nourishment for the developing embryo as a haploid, female gametophyte that protects, preconditions, and enhances proembryo development for future precise environmental adaptation;

4) new directions in the diagnosis of "adaptive phenotypes" and how these relate to current trends in forestry practices.

More specifically, I will attempt to show that conifers adapt to water-related climatic stress by:

1) polyembryony of various types, as revealed by studies with artificial ovules and bioreactors,

2) the immediate assimilation of water into critical metabolic pathways and into structures, such as DNA in chromatin,

3) metabolic thresholds in the new generation that affect i) replicator activities associated with the cell cycle and ii) opportunistic histogenic algorithms in proembryony that provide alternative patterns in development, and

4) mother tree influences that physiologically precondition the new generation to climate at the seed source.

Since many gaps remain in establishing the above observations, I can only speculate on future trends relating climatic change to biotechnologic opportunities in conifer polyembryony.

CLIMATE DETERMINES THE RELEASE OF POLLEN
AND LEADS TO INTROGRESSION

The warmer climates at the southern limits of the natural boreal range tend to release pollen earlier. This pollen is carried by the upward air currents (trade winds) and is deposited on female reproductive cones that are just opening in northern regions. If fertilization occurs, this introgression, together with self-pollination at a later date, can lead to genetically inferior and poorly adapted seeds for northern ranges. Self-pollination also leads to inbreeding depression (Fowler, 1978). Seed breeding and production orchards are susceptible to this flow of pollen.

Orchards must be strategically situated and controlled. Some industries actually release cold water from aircraft to retard the release of pollen from southern sources and to maintain better genetic control of the quality of seed. Other seed orchard strategies based more on local conditions have been described by Wright (1976).

CLIMATIC AND CULTURAL EFFECTS ON POLYEMBRYONY

Our next concern is with seed development in the female reproductive cones after fertilization. The mother tree provides, through the vascular connections, the nutrition and water to the ovuliferous scale. In many instances, a single ovule can produce many eggs. In Douglas fir from Oregon and Washington, I have seen up to seven eggs in one ovule. When these are fertilized, the new generation is not clonal, but represents fraternal products. This type of polyembryony has been called "simple" or "polyzygotic". Each egg is a genetically distinct meiotic reduction-product of the mother tree. Added variation comes from pollen from the same tree (inbreeding) or from outside sources. I will not deal further with this mode of ensuring greater numbers of surviving offspring. Reference can be made to Dogra (1980, 1983), Dawkins (1982), and Klekowski (1988).

The University of California has licensed to industry a novel cloning process based on somatic polyembryony with rescued embryonal-suspensor masses. The process enables the potential mass production of embryos and artificial seeds from elite germplasm sources. The technology allows for the imposition of local and specific environmental fixes. It takes advantage of the local genetic diversity to multiply embryos by somatic polyembryony throughout the year and independently of the vagaries of nature. Cell lines for the process have been established on a laboratory scale as cell suspension cultures. These cells, which have been maintained for nearly seven years, represent some of the most advanced and genetically improved trees in Europe and North America. Our Norway spruce lines (*Picea abies*) and Douglas fir (*Pseudotsuga menziesii*) were obtained from Kemira Oy (Finland) and Weyerhaeuser (Tacoma, WA) or the State of California (Davis, CA), respectively. Weyerhaeuser has licensed the process for two years and is ready for large-scale field tests to study genotype x environmental interactions for the clonal materials.

In somatic polyembryony, multiple proembryos are "reconstituted" from the zygotic proembryo by a cleavage process which has several variations *in situ* (Dogra, 1967, 1980, 1983) and in vitro (Durzan, 1988a & 1988b; Durzan & Durzan, 1990). Here the reconstituted products are clonal. In nature, multiple embryos are normally produced in response to climatic and environmental stress. Stresses in the mother tree, as reflected by embolism and cavitation, could also elicit the cleavage process in the ovule, although it is not long before the early developing seed becomes independent of the immediate influences of the mother tree. Because the cleavage products are genetically similar, they offer little selective advantage in genetic terms. Multiplication helps, however, to ensure that at least one embryo survives the stressful episodes. Normally only one embryo develops as the seed ripens. At an early stage, proembryos release considerable mucilage to digest away the maternal nutrient reserves. The mucilage forms a plug in the erosion zone and releases products that inhibit the growth of other proembryos. The viscosity of this mucilage is a sometime function of water availability from the air or mother tree and of enzyme activity.

The chance of recovery of multiple embryos from a single ripe seed increases as arctic regions are approached (Simak, 1973; Sziklai, 1986). Furthermore, proembryonal cell lines, used for cloning, can be cryopreserved in liquid nitrogen with retention of morphogenic potential after thawing (Gupta, Durzan & Finkle, 1987).

Other multiplicative mechanisms reflect attempts for survival and lead to aberrant development. These often occur in response to drought and cold. The tendency for conifers to amplify processes for adaptation and survival is a characteristic evolutionary feature. The tendency is seen as:
1) a geometric increase in the amount of DNA in the genome (Price, Sparrow & Nauman, 1973),
2) a flexible 4-tiered proembryo originating from a two-cell basal plan, which can geometrically reconstitute proembryos by the cleavage and budding process (Dogra, 1967, 1980),
3) the geometric production of multiple clonal embryos from one fertilization event by the cleavage and budding process (Dogra, 1967, 1980; Durzan 1988b), and
4) the formation of double or multiple female gametophytes in a seed (Simak, 1973).

The relationship of the amplified geometric replications to environmental and climatic stress needs further investigation. This investigation may be seen in terms of opportunistic developmental outcomes fostering adaptation and survival of the genotype. Such studies can now be explored under more controlled environmental conditions.

BIOREACTOR SCALE-UP OF EMBRYONAL-SUSPENSOR MASSES USING CLIMATIC VARIABLES AND SPECIFIC SEED SOURCES

With elite and well-adapted trees, we can rescue from each ovule one set of cleavage products from each egg. This is best done during the first six to seven weeks after fertilization; otherwise, the rescued products tend to produce callus (Hong *et al.*, 1991). The rescued products are components of the embryonal-suspensor mass (ESM), (Durzan & Gupta, 1988). In this mass, there is a developmental gradient from the most developed embryo to yet undeveloped proembryonal cells. Visual selection and culture of the undeveloped cells in darkness easily leads to a scale-up of cell lines in 1-liter nippled flasks. The culture system is analogous to an artificial ovule, preloaded with nutrients that enable proembryonal cells to reconstitute new proembryos without callus formation.

A number of elite genotypes of Norway spruce (*Picea abies* Karst.) from Finland have been rescued as ESMs (Jokinen & Durzan, unpublished observations, 1991). Cell suspension cultures were established in bioreactors (Santerre & Durzan, 1990). During batch cultures, ESMs revealed a number of interesting hydrodynamic properties (Durzan, 1990a) which may contribute to the cleavage process and to other stress-induced phenomena (Durzan, 1990a; Durzan, 1991).

The multiplication of proembryonal cells occurs by at least two processes: 1) cleavage and 2) budding of daughters of proembryonal cells (*viz.* embryonal tube, embryonal suspensor cells). Often, when proembryos cleave, they remain stuck together, and this creates a problem for further development. The cleavage process can easily be blocked by the addition of abscisic acid. Reduction in the level of plant growth regulators selects for individual embryos that are able to continue development on their own (*e.g.* Boulay, Gupta & Durzan, 1988; Hong *et al.*,1992). Abscisic acid is a metabolite produced under water stress which may lead to abscission of leaves (Li & Walton, 1990). In ESMs of conifers, abscisic acid not only blocks polyembryony, but induces the formation of protein bodies (Roberts *et al.*, 1990).

Too much abscisic acid will create later problems by inhibiting germination (Boulay *et al.*, 1988). We do not yet know how water stress and cavitation in the mother tree can affect the production of abscisic acid around the ESM in a developing seed. In field-grown maize, differences in abscisic acid content in aborting and non-aborting kernels were observed only after embryo abortion was complete (Reed & Singletary,

1989).

In budding polyembryony, development is reset by the cell cycle to the point just after fertilization by an initial free nuclear stage. Nuclei migrate to opposite poles of the budding cell to reconstitute the basal plan of proembryo development, viz. a primary proembryonal and an upper suspensor cell: pE and pU in Dogra's terminology (1980).

The recovery of budding cells with free nuclei can be confused with "necking" products induced by osmotic stress (Durzan, 1991). The necking products are different-sized protoplasts within the cell. Some protoplasts contain trapped nuclei and others may be rich in organelles. Necking occurs mainly in cylindrical cells showing rapid cytoplasmic streaming. Necking is due to instabilities of flowing protoplasmic masses. Instability is caused by adjustment to osmotic and gas pressures. This disruptive physico-chemical process contrasts with cavitation and embolism, which disrupt water and nutrient movement in the mother tree (cf. Milburn, this volume). In necking, the residual effect is the fragmentation of protoplasm. We do not know how the water status in the vascular elements of the mother tree, connected to the developing seed, affects budding polyembryony or if water stress can induce necking in cells of the ESM. The phenomenon of budding can be studied under laboratory conditions by removal of cells from artificial ovules for microscopic study, much like that for the demonstration of cavitation (cf. Milburn, this volume).

Climatic stress is suspect in affecting the size, quality and preconditioning of embryos in developing seeds, as seen by correlations between seed parameters and local climate at the seed source (Table 2, cf. Durzan & Chalupa, 1968). These parameters are now being built into the design of model-reference "process controls" for bioreactors to precondition clonally produced embryos targeted for northern climates (Durzan & Durzan, 1990).

Climate-related problems in seed development have a bearing on the classification and interpretation of the "penetrance" and expression of lethal genes introduced by selfing and introgression. From X-ray studies of seeds, several developmental abnormalities in embryo and gametophyte development are known to characterize lethality (Simak, 1973). These environmentally-induced responses are called "conditional lethals" (cf. Rieger, Michaelis & Green, 1976). For any given genotype, climate contributes to developmentally related phenotypic variances that need to be sorted out during the evaluation of clonal and genotypic performance (Durzan, 1989).

Artificial seeds can be made by encapsulating somatic embryos that are reconstituted by somatic polyembryony (Durzan & Gupta, 1988; Gupta & Durzan, 1987). Before climate at the seed source can be evaluated for these artificial seeds, baseline data are needed to understand how zygotic seeds perform under similar conditions.

METABOLISM OF TRITIATED WATER
AND THE PROEMBRYONAL CELL CYCLE

Based on the earlier climatic correlations of Durzan and Chalupa (1968) (cf. Table 2), a series of studies were taken with jack pine seeds from a wide range of northern Canadian seed sources with the aim of finding:
1) the first covalently recovered product of tritiated water in the soluble and insoluble fraction of initially dry seeds,
2) the role of light and darkness as a method for controlling the inhibition and germination of seeds in response to tritiated water, and
3) isotopic effects that lead to the deterioration of the processes being studied.

Seeds, freshly removed from reproductive cones, showed variable germination in darkness. Variability could be reduced to zero by preexposure of seeds to a few minutes of far-red light (Campbell & Durzan, 1979). The addition of water to dry seeds and exposure to red light was quickly followed by 99-100% germination. Seeds were precleaned and sorted to remove lethals and empty seeds from the seed lot, so as not to bias experimental results.

For one of the seed sources (Petawawa Plains, Ontario, Canada), tritium, fed to dry seeds as tritiated water, could be recovered in a covalent form from the 80% (v/v) alcohol-soluble and -insoluble fractions (Durzan, 1983; Durzan, Mia & Wang, 1971). In the alcohol-soluble fraction, covalently bound tritium was recovered within hours, almost entirely from glucose and fructose. This indicated the hydrolysis of sucrose and starch and the possible entry of tritium into glucose by intramolecular hemiacetal formation. By high resolution autoradiography of cells, we could localize most of the cytoplasmic radioactivity over starch grains in amyloplasts. A few hours later, radioactivity could be recovered from the soluble, amino acid fraction. Proline, glutamic acid and alanine were the main metabolites containing covalently bound tritium.

Proline usually accumulates during water stress and becomes labelled through the specific incorporation of tritium linked to the reduction of oxidized pyridine nucleotides. Tritium, from tritiated water, enters intermediary metabolism with the reduction of pyridine nucleotides. Next, through enzymatic action, the tritiated nucleotides transfer their tritium to covalent positions in a variety of substrates. By recovery of these metabolites, the dominant enzymatic reactions occurring at that time were revealed. Prior heat-killing of seeds as "controls" did not lead to recovery of any covalently labelled substrates.

TABLE 2. Equations showing the linear regression of composition of the female gametophyte and embryo on climatic factors, seed collection data, and composition. The coefficient of determination ($100R^2$) gives the percentage reduction in sum of squares of the dependent variable.* See Durzan and Chalupa (1968) for units upon which the parameters X and Y are based.

Y	= a +/- b	(X)	Error d.f.	$100R^2$
Female Gametophyte				
Dry weight	= 123.2 + 0.03058	(degree-days)	13	49.7
Per cent soluble sugar	= 2.270 + 2.760 x 10^{-3}	(photoperiod)	10	76.3
Per cent stachyose	= 1.261 + 1.433 x 10^{-3}	(photoperiod)	10	71.0
Per cent sucrose	= 0.5081 + 8.601 x 10^{-4}	(photoperiod)	10	72.5
Free amino acids	= 1219 - 1.085 x 10^{-3}	(degree-days x dry weight)	6	70.6
Arginine	= 346.5 - 0.01566	(temperature x precipitation)	6	60.1
Amides	= 32.32 + 0.02680	(precipitation)	6	44.1
Soluble protein	= 20.23 - 4.567 x 10^{-6}	(degree-days x dry weight)	10	79.1
Soluble protein	= 15.09 + 4.250 x 10^{-3}	(free amino acids)	6	72.0
Embryo				
Dry weight	= 9.921 + 8.472 x 10^{-3}	(degree-days)	13	69.6
Embryo length	= 2.066 + 2.499 x 10^{-4}	(degree-days)	13	61.3
Per cent soluble sugar	= 6.044 + 6.753 x 10^{-3}	(photoperiod)	10	67.2
Per cent stachyose	= 5.295 - 4.017 x 10^{-7}	(degree-days x precipitation)	10	70.1
Per cent sucrose	= 1.555 + 2.359 x 10^{-3}	(photoperiod)	10	60.4
Free amino acids	= 6313 - 0.04589	(precipitation x dry weight)	7	66.4
Arginine	= 2493 - 0.1199	(temperature x precipitation)	7	73.0
Amides	= 3103 - 73.19	(soluble proteins)	7	55.5
Alanine	= 426.3 - 0.01934	(temperature x precipitation)	7	63.8
Soluble protein	= 42.37 - 0.1609	(dry weight)	10	78.9
Soluble protein	= 33.51 + 0.001815	(arginine)	7	72.7

*Stepwise regression

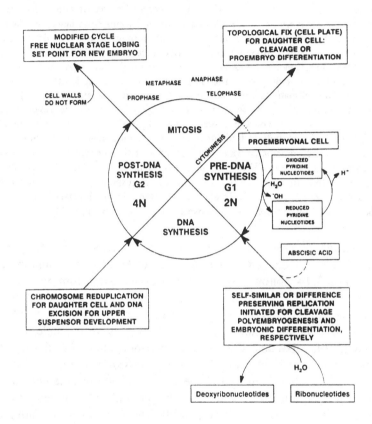

FIG. 1. The cell cycle and polyembryogenesis. Events in a proembryonal cell cycle determine alternative pathways leading to cleavage and budding polyembryony or to the differentiation of the proembryo cell into a single embryo. In cleavage, DNA replication during the S phase requires that a self-similar code be replicated to reconstitute multiple identical proembryos. In embryonic differentiation, the daughter cells are programmed by difference-preserving replication events and sites during the onset of the S phase. Once DNA synthesis is complete in the S phase, chromatid and chromosome reduplication are completed during G2 where the level of DNA becomes 4N. If a cell wall is not formed during mitosis, a free-nuclear stage appears and polar nuclear migration can reset development back to the original 2-cell basal plan characteristic for all gymnosperms. If a wall is formed, then the daughter cell's position is indicated by the angle taken by the cell plate. The daughter cell can be either another identical proembryonal cell (polyembryogenesis) or the next cell involved in the formation of the characteristic tier of proembryonal cells that is associated with the development of an individual embryo (embryogenessis).

FIG. 1 (continued) Other cells derived from one of the two basal plan cells produced after nuclear migration (free nuclear stage) contribute to the development of the upper suspensor system. It is thought that DNA excision might occur in this case.The difference in replication events contributing to polyembryony or embryony can be followed by the recovery of covalently bound tritiated water from newly synthesized DNA at the S phase. Tritium enters DNA covalently through ribonucleotide reductase, which converts the pool of ribonucleotides to ribonucleotides for DNA synthesis. "Hot spots" or Okazaki fragments representing the replication sites of DNA can be detected by high resolution autoradiography.The G1 phase of the cell cycle is most sensitive to environmental and climatic factors. During heat or drought, abscisic acid (ABA) can be formed, which may inhibit the formation of multiple embryos by cleavage. ABA is used in bioreactors to encourage rescued proembryonal-suspensor masses to stop polyembryogenesis and initiate individual embryo development (*e.g.* Boulay *et al.,* 1988). The tritium from tritiated water can be used to reduce enzymatic cofactors such as the pyridine nucleotides. The reduced (tritiated) nucleotides can then covalently transfer their tritium to substrates that are produced in response to the climatic stress, e.g. proline (Durzan *et al.,* 1971). In other cases, starch in amyloplasts is hydrolyzed with the incorporation of tritium into glucose and fructose (Durzan, 1983).

The structural components of cells were studied next. In nuclei, radioactivity was localized over replication forks of the chromatin. The extraction of DNA, followed by recovery of covalent tritium by chemical degradation of DNA, showed that most of the tritium was found in the deoxyribose moiety (Durzan, 1983). This implicated an important role for the enzyme "ribonucleotide reductase", which is responsible for the conversion of ribonucleotides to deoxyribonucleotides during the synthesis of DNA (Reichard & Ehrenberg, 1983). The entry of tritium into DNA could also occur by linkage with reduced pyridine nucleotides that transfer tritium from water to thioredoxin or glutaredoxin via ribonucleotide reductase (cf. Stryer, 1988). These events were among the first observed reaction-products of tritiated water. During imbibition, the cell cycle becomes reactivated with most cells initially at G1, passing to the S or DNA synthetic phase in waves of mitotic activity (Fig. 1).

While considerable insight can be obtained through isotopic studies with tritiated water, due caution must be given to isotopic effects. To avoid the lethal effects of tritium, we could not exceed doses of 10 μCi $ml^{-1} g^{-1}$ of seed.

To summarize, our results show 1) how water metabolism relates directly to composition and to structural changes at the nuclear genome level of individual cells, and 2) how a few critical enzyme systems may be employed by the pioneering habit of trees in their struggle to survive harsh conditions. These opportunistic metabolic patterns somehow permit the staggered multiplication of proembryos that enable the new generation to survive. We do not know yet how directly the enzymatic systems linked to the assimilation of water are affected by acid deposition in forests (Schulze, 1989) or to climatic change (Layser, 1980).

FUTURE CLIMATIC CHANGE AND POLYEMBRYONY

Undoubtedly, future climatic changes will impact on how trees reproduce in climatically sensitive zones and in seed orchards. Now, local environmental variables can be simulated and imposed on selected genotypes under bioreactor conditions. This approach should be paralleled with studies on mother trees under orchard conditions. In Douglas fir under field conditions, Waring and Running (1978) have shown that cavitation occurs when polyembryony would be expected.

New methods are available to characterize how ESMs respond to a variety of environmental factors. Response surfaces, based on dynamic profiles of nitrogen metabolic flux in organs of trees in relation to local, present and past microenvironmental conditions, are well-suited for this task (Durzan, 1987).

From an examination of the response surfaces, we can attempt to find evidence for self-adjustment (self-optimization) in the face of changing diurnal and seasonal environments around the mother tree. Can the environmental fluctuations program recognizable, adaptive events into the proembryos? How does the genotype affect the response of hierarchical metabolic phenotypes to a specific external change? Are adaptive metabolic phenotypes self-designing or self-demultiplexing? Can metabolic phenotypes repair and restore themselves, *i.e.* can they adjust to certain kinds of internal defects?

This type of "response mapping" is a logical outcome of a confluence of developments. These are: past metabolic studies with the phenology of fruit and forest trees (Durzan, 1971, 1987, 1988a, 1989, 1990b), systems ecology (Kitching, 1983), multiple criteria for climatic and atmospheric changes (*e.g.* remote sensing) (Layser, 1980; Sabins, 1987), the advent of computer-assisted technologies for pattern (Fu, 1982), threshold logic (Dertouzos, 1965), dynamic systems (Demongeot, Golès & Tschuente, 1985), mathematics of fractals and chaos (Mandelbrot, 1983; Gleick, 1987), and coding theory (Hill, 1986).

A start in characterizing the behavior of fruit trees has also been made. Satellite images of an orchard under stress can readily be obtained. These images show leaf temperature differences in a color-coded topography. This information on tree behavior can be correlated with response surfaces for metabolic flux in leaves and fruits.

In all trees, signal processing is complicated by physiological preconditioning, maturation, senescence and aging (Durzan, 1990c). Preliminary results with leaves and buds of almond trees, with and without bud failure, also indicate that the flux of soluble nitrogen of organs reveals hidden, metabolic lesions associated with bud failure responses under field conditions (Kester, Durzan & Shackel, 1990). This approach can also be used to assess the adaptability and competitiveness of transgenic trees when placed in the natural and artificial environment.

Once the dynamic behavior of climate and tree adaptability are established, models for the study of artificial life (Langton, 1989) and applications of expert systems (Kline & Dollins, 1989) should enable the scientist to simulate future changes. The problem here becomes the scientist's dependency upon model-driven rather than data-driven reasoning. Moreover, most simulations could not be easily duplicated by an experimental field situation. Even so, the spectrum of clues obtained from simulations could greatly assist in the design of simpler field experiments to evaluate phenotypes against different genetic and clonal backgrounds. This approach has been termed "molecular phenogenetics" (Durzan, 1990b).

Another direction for future research is the experimental restoration of earlier climatic variables associated with the "evolution of the seed habit". My laboratory is currently collecting data on Paleozoic and Mesozoic climatic variables that could be imposed on cell culture systems. This may enable us to observe how a wide range of variables modify the opportunistic degrees of freedom in polyembryony and show how metabolic adaptation occurs during evolution under harsh conditions.

LIMITS ON ADAPTATION EVOLUTION

If adaptation were dominant, every conifer would mature rapidly, reproduce at an infinite rate and live forever. No trees meet all of these criteria, although some are very prolific and long-lived. Constraints that limit adaptation and development are sought at the genetic level because selection (directional, stabilizing, etc.) cannot produce evolutionary change if genetic variation is lacking (Loeschske, 1987). This type of reasoning leads to an assessment of how patterns of genetic variation limit polyembryony and adaptive evolution. Hence, the sources of genetic constraints on the life history of conifers become important. In this context, the actions of pleiotropy (*e.g.* one gene affects many traits), polymery (*e.g.* several genes affect one trait) and genome rearrangements might define constraints to adaptation and the expression of adaptive phenes under strong directional selection (*e.g.* Yablokov, 1986). Little is known about the structure of genetic correlations under these conditions (Li & Graur, 1991).

POLYEMBRYONY IN ZYGOTIC, CLONAL AND FAMILY FORESTRY

"Zygotic forestry" (cf. Libby, 1992) involves the deployment of seeds and seedlings. Some of the limitations of this approach are:
1) uncertainty of genotypes due to crosses and subsequent polyembryony, *i.e.* poorly characterized or not well-known genotypes.
2) performance knowledge must focus on field testing of progeny as they relate to family, line, breed or population.
3) management cannot yet take advantage of differences among individual clones or of extensive knowledge about a particular clone.

By recovery of all products of polyembryony in a single developing seed, the uncertainty around the rescued genotypes can now be studied and evaluated. Furthermore, all products of controlled or uncontrolled crosses can be cloned and scaled-up *en masse* for field testing at different sites. These genotypes can also be studied for their adaptability to specific environmental gradients, sequences and sites.

The phenotypic variances that arise from somatic polyembryony need to be sorted out (Durzan, 1989). Greater management of clonal and family genotypic and phenotypic variances may be possible in terms of time-saving embryonal-juvenile-mature correlations and accelerated testing programs using the tools of molecular phenogenetics.

Somatic polyembryony now presents "zygotic forestry" with new opportunities for rapid, vegetative multiplication and improved assessment. Other classical forms of propagation are dependent upon the restoration of missing parts (rooting) or on micropropagation and adventitious bud multiplication.

"Clonal forestry" involves the deployment of relatively few well-tested and well-understood clones to increase management efficiency and effectiveness (Libby, 1991). It also allows the deployment of clones that would rarely survive selection in genetically heterogeneous mixtures or that would rarely occur.

"Family forestry" involves the collection of open pollinated (orchard), polycross and/or full-sib families for separation and deployment as single or prescribed mixes (Sweet & Krugman, 1977; Carson 1986). In this approach, good progeny-test data are needed.

Climate at the seed source will more or less affect the performance of seeds under all types of forestry deployments. The role of water at the critical stages of seed development in preconditioning seeds and directing opportunistic survival mechanisms has been indicated.

REFERENCES

Boulay, M.P., Gupta, P.K. & Durzan, D.J. (1988) Conversion of somatic embryos from cell suspension culture of Norway spruce (*Picea abies* Karst.). *Plant Cell Reports*, 7, 134-137.

Campbell, R.A. & Durzan, D.J. (1979) Laser activation of phytochrome-controlled germination in *Pinus banksiana. Canadian Journal of Forest Research*, 9, 522-524.

Carson, M.J. (1986) Advantages of clonal forestry for *Pinus radiata* - real or imagined? *New Zealand Journal of Forest Science*, 16, 403-415.

Dawkins, R. (1982) *The Extended Phenotype*. W.H. Freeman, San Francisco.

Demongeot, J., Golès, E. & Tschuente, M. (eds.) (1985) *Dynamic Systems and Cellular Automata*. Academic Press, New York.

Department of Energy/Energy Research. (1989) *Atmospheric carbon dioxide and the greenhouse effect*. National Technical Information Service, U.S. Dept. Commerce, Springfield, VA, NTIS-PR-360.

Dertouzos, M.L. (1965) *Threshold Logic: A Synthesis Approach*. MIT Press, Cambridge, Massachusetts.

Dogra, P.D. (1967) Seed sterility and disturbances in embryogeny in conifers with particular reference to seed testing and tree breeding in Pinaceae. *Studia Forestalia Suecica*, 45, 1-97.

Dogra, P.D. (1980) Embryogeny of gymnosperms and taxonomy -- an assessment. In *Glimpses in Plant Research* (ed. P.K.K. Nair), vol. 5, pp. 114-128. Vikas House Limited, New Delhi.

Dogra, P.D. (1983) Reproductive biology of conifers and its application in forestry and forest genetics. *Phytomorphology*, 33, 142-156.

Durzan, D.J. (1971) Free amino acids as affected by light intensity and the relation of responses to the shade-tolerance of white spruce and shade intolerance of jack pine. *Canadian Journal of Forest Research*, 1, 131-140.

Durzan, D.J. (1983) Metabolism of tritiated water during imbibition and germination of jack pine seeds. *Canadian Journal of Forest Research*, 13, 1204-1218.

Durzan, D.J. (1987) Physiological states and metabolic phenotypes. In *Cell and Tissue Culture in Forestry* (ed. J.M. Bonga & D.J. Durzan), 2nd Edition, Volume 2, *Principles and Methods*. pp. 405-439. Martinus Nijhoff/Dr. W. Junk.

Durzan, D.J. (1988a) Plant growth regulators in cell and tissue culture of woody perennials. *Journal of Plant Growth Regulation*, 6, 95-111.

Durzan, D.J. (1988b) Process control in somatic polyembryogenesis. In *Molecular Genetics of Forest Trees* (ed. J.E. Hällgren), pp. 147-186. Frans Kempe Symposium, Swedish University of Agricultural Science, Umea, Sweden.

Durzan, D.J. (1989) Performance criteria in response surfaces for metabolic phenotypes of clonally propagated woody perennials. In *Application of Plant Biotechnology in Forestry* (ed. V. Dhawan), pp. 181-203. Plenum Press, New York.

Durzan, D.J. (1990a) Physiological aspects of somatic polyembryogenesis in suspension cultures of conifers. In *International Symposium on Forest Tree Physiology. Annales des Sciences Forestières,* **46 (Suppl.),** 101-107.

Durzan, D.J. (1990b) Molecular phenogenetics as an aid to fruit breeding. *Proceedings of the International Symposium In Vitro Culture and Horticultural Breeding.* May 30 - June 3, 1989. Cesena, Italy (ed. J. Janich & R.H. Zimmerman). *Acta Horticulturae,* **280,** 547-556.

Durzan, D.J. (1990c) Genetic manipulation of forest trees: Implications for physiological processes. In *Plant Aging: Basic and Applied Approaches* (ed. R. Rodriguez, R. Sánchez Tamés & D.J. Durzan), pp. 303-309. Plenum Press, New York.

Durzan, D.J. (1991) Necking in embryonal tube cells and its implications for morphogenic protoplasts and tree improvement. In *Woody Plant Biotechnology* (ed. M.R.Ahuja), pp 123-137, Plenum Press, New York.

Durzan, D.J. & Chalupa, V. (1968) Free sugars, amino acids, and soluble proteins in the embryo and female gametophyte of jack pine related to climate at the seed source. *Canadian Journal of Botany,* **46,** 417-428.

Durzan, D.J. & Durzan, P.E. (1990) Future technologies: Model-reference control systems for the scale-up of embryogenesis and polyembryogenesis in cell suspension cultures. In *Micropropagation* (ed. P. Debergh & R.H. Zimmerman), pp. 389-423. Kluwer Academic Publishers.

Durzan, D.J. & Gupta, P.K. (1988) Somatic embryogenesis and polyembryogenesis in conifers. Downstream processes. *Advances in Biotech Processes,***9,** 53-81.

Durzan, D.J., Mia, A.J. & Wang, B.S.P. (1971) Effects of tritiated water on the metabolism and germination of jack pine seeds. *Canadian Journal of Botany,* **49,** 2139-2149.

FAO, Food and Agriculture Organization of the United Nations (1985) *Tropical Forestry Action Plan.* Committee on Forest Development in the Tropics, Rome, Italy.

Fowler, D.P. (1978) Population improvement and hybridization. *Unasylva,* **30,** 21-26.

Fu, K.S. (1982) *Syntactic Pattern Recognition and Applications.* Prentice-Hall Inc., Englewood Cliffs, New Jersey.

Gleick, J. (1987) *Chaos. Making a New Science.* Viking Penguin Inc., New York.

Gupta, P.K. & Durzan, D.J. (1987) Biotechnology of conifer-type somatic poly-embryogenesis and plantlet regeneration in loblolly pine. *Bio/Technology,* **5,** 147-151.

Gupta, P.K., Durzan, D.J., & Finkle, B.J. (1987) Somatic polyembryogenesis in embryogenic cell masses of *Picea abies* (Norway spruce) and *Pinus taeda* after thawing from liquid nitrogen. *Canadian Journal of Forest Research,* **17,** 1130-1134.

Hill, K. (1986) *A First Course in Coding Theory.* Clarendon Press, Oxford.

Hong, L., Boulay, M., Gupta, P.K. & Durzan, D.J. (1992) Variations in somatic polyembryogenesis: induction of adventitious embryonal-suspensor masses on developing Douglas fir embryos. In *Woody Plant Biotechnology* (ed. M.R. Ahuja), pp 105-121, Plenum Press, New York.

Houghton, J.T. (ed.). (1984) *The Global Climate.* Cambridge University Press, Cambridge.

Kester, D.E., Durzan, D.J. & Shackel, K. (1990) Relation between phenotypic expression of noninfectious bud-failure (NBP) in almond and the seasonal growth-dormancy-rest cycle in almond shoots. *Proceedings of XXIII International Horticultural Congress,* Florence, Italy. Abstract 2226.

Kitching, R.L. (1983) *Systems Ecology.* University of Queensland Press, St. Lucia, Queensland, Australia.

Klekowski, E.J., Jr. (1988) *Mutation, Developmental Selection, and Plant Evolution.* Columbia University Press, New York.

Kline, P.J. & Dollins, S.B. (1989) *Designing Expert Systems.* John Wiley and Sons, New York.

Langton, C.G. (ed.). (1989) *Artificial Life.* Addison-Wesley Publishing Company, Redwood City, California.

Layser, E.F. (1980) Forestry and climatic change. *Journal of Forestry,* **78,** 678-682.

Li, W.-H. & Graur, D. (1991) *Fundamentals of Molecular Evolution.* Sinauer Associates, Inc., Sunderland, Mass.

Li, Y. & Walton, D.C. (1990) Violaxanthin is an abscisic acid precursor in water-stressed dark-grown bean leaves. *Plant Physiology,* **92,** 551-559.

Libby, W.J. (1992) Use of genetic variation in breeding forest trees. In *Plant Breeding in the 1990s* (eds. H.T. Stalker & J.P. Murphy), pp 101-117, CAB Internationsl.

Loeschske, V. (ed.). (1987) *Genetic Constraints on Adaptive Evolution.* Springer-Verlag, New York.

Mandelbrot, B.B. (1983) *The Fractal Geometry of Nature.* W.H. Freeman & Company, New York.

Price, J.H., Sparrow, A.H. & Nauman, A.F. (1973) Evolutionary and developmental considerations of the variability of nuclear parameters in higher plants. I. Genome volume, interphase chromosome volume and estimated DNA contents of 236 gymnosperms. In *Basic Mechanisms in Plant Morphogenesis,* pp. 390-421. Brookhaven National Lab Symposium, No. 25, June 4-6.

Reed, A.J. & Singletary, G.W. (1989) Roles of carbohydrate supply and phyto-hormones in maize kernel abortion. *Plant Physiology,* **91,** 986-992.

Reichard, P. & Ehrenberg, A. (1983) Ribonucleotide reductase - A radical enzyme. *Science,* **221,** 514-519.

Rieger, R., Michaelis, A. & Green, M. (1976) *Glossary of Genetics and Cytogenetics.* Springer-Verlag, New York.

Roberts, D.R., Flinn, B.S., Webb, D.T., Webster, F.B. & Sutton, B.C.S. (1989) Abscisic acid and indole-3-butyric acid regulation of maturation and accumulation of storage protein in somatic embryos of interior spruce. *Physiologia Plantarum,* **78,** 355-360.

Roberts, D.R., Sutton, B.C.S. & Flinn, B.S. (1990) Synchronous and high frequency germination of interior spruce somatic embryos following partial drying at high relative humidity. *Canadian Journal of Botany,* **68,** 1086-1090.

Sabins, F.F,Jr. (1987) *Remote Sensing. Principles and Interpretation.* W.H. Freeman

and Co., New York.
Santerre, A. & Durzan, D.J. (1990) Etude dynamique du métabolisme azote d'un système cellulaire polyembryonique cultivé in vitro. *Proceedings Second Forum for Scientific Exchange Consulat Général de France,* SUNY, Stony Brook, June 8 to 11.
Schulze, E.-D. (1989) Air pollution and forest decline in a spruce *(Picea abies)* forest. *Science,* **244,** 776-783.
Silen, R.R. (1982) Nitrogen, corn, and forest genetics. The agricultural yield strategy - Implications for Douglas-fir management. USDA Forest Service General Technical Report PNW-137.
Simak, M. (1973) Polyembryonal seeds of *Pinus silvestris* in arctic regions. Royal College Forestry Stockholm Research, Note No. 45, 1-14.
Stryer, L. (1988) *Biochemistry,* 3rd edn. W.H. Freeman Co., New York.
Sweet, G.B. & Krugman, S.L. (1977) Flowering and seed production problems - and a new concept of seed orchards. In *Proceedings of Third World Consultation of Forest Tree Breeding,* pp. 749-759, Canberra.
Sziklai, O. (1986) Polyembryony of *Pinus contorta* Doug. in central Yukon. In *Provenances and Forest Tree Breeding for High Latitudes* (ed. D. Lindgren), Report No. 6, Swedish University of Agricultural Sciences, Umea.
Waring, R.H. & Running, S.W. (1978) Sapwood water storage: Its contribution to transportation and effect upon water conductance through the stems of old-growth Douglas-fir. *Plant and Cell Environment,* 1, 131-140.
Wright, J.W. (1976) *Introduction to Forest Genetics.* Academic Press, New York.
Yablokov, A.V. (1986) *Phenetics, Evolution, Population, Trait.* Columbia University Press, New York.
Yapp, C.J. & Epstein, S. (1982) Climatic significance of the hydrogen isotope ratios in tree cellulose. *Nature,* **297,** 636-639.

A heat balance method for measuring sap flow in small trees

C. VALANCOGNE AND Z. NASR

Laboratoire de Bioclimatologie,
INRA, Centre de Bordeaux,
Domaine de la Grande Ferrade, BP 81
F3388 Villenave d'Ornon, France.

SUMMARY

The heat balance method of Sakuratani (1981) was adapted and improved for measuring sap flow in fruit trees of high density orchards. In this technique the trunk is surrounded by a heating jacket operating at constant power. Thermocouples are used to measure conductive and convective heat flux and therefore the sap flow rate from the heat balance. This method does not require any calibration before the setting up of the sap flow sensor.

INTRODUCTION

Several thermal methods have been used for measuring sap flow in the trunk of trees: heat pulse methods (Marshall, 1958; Swanson & Whitfield, 1981) and those that depend on the temperature of a linear heat source inserted in the xylem (Granier, 1985). These methods give the sap flux density (kg s^{-1} m^{-2}). The estimation of the sap flow rate (kg s^{-1}) involves the measurement of the cross-sectional area of the functional xylem. An improvement of the heat pulse method (Cohen, Fuchs & Green, 1981) avoids this additional measurement by estimating the radial profile of sap velocity, but assembling the sensor is very difficult. Heat balance methods were applied also on sectors of tree trunks by Daum (1967), Cermák, n, Kucera & Penka (1976). With all these methods, there is a question of how representative the measurements are of the overall flow rate in the trunk.

Here we propose a direct measurement of the sap flow rate in the whole section of the stems or trunks as an improvement of the heat balance method (Sakuratani, 1981; Valancogne & Nasr, 1989a, 1989b).

MATERIALS AND METHODS

A heating jacket whose length L (m) is one or two times that of the stem diameter is fitted around the stem and delivers a constant heat power W (W) (Fig. 1).

Different heat fluxes are delivered by the heating jacket:
i) a convective heat transmission by the sap, q_{sap} (W):

$$q_{sap} = c_w \, d_s \, \delta T \qquad \text{Equation 1}$$

where c_w is the heat capacity of water, 4180 J kg^{-1} $^\circ$C^{-1}, δT the temperature elevation ($^\circ$C) of the sap across the heated segment of stem and d_s the sap flow rate in the stem (kg s^{-1});
ii) upstream and downstream conductive heat flows, q_u and q_d (W);
iii) a lateral heat flux, q_{lat} (W).

There is also the rate of heat stored q_{sto} (W) in the volume delimited by the sensor according to the variation of its temperature.

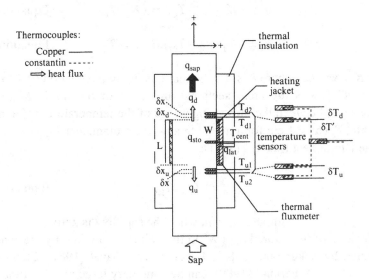

FIG. 1. Design of the heat balance sap flow sensor and temperature sensors.

Therefore, out of the heat balance of the heated segment is derived the heat flow carried by the sap, q_{sap} (W):

$$q_{sap} = W - (q_d - q_u) - q_{lat} - q_{sto} \qquad \text{Equation 2}$$

and the sap flow, d_s (kg s^{-1}) is computed as :

$$d_s = q_{sap} / (c_w \, \delta T) \qquad \text{Equation 3}$$

The heating jacket is built from a constantan wire fixed onto adhesive tape. For computing the conductive and convective heat fluxes, two pairs of copper cylinders enclosing a copper-constantan thermocouple are radially inserted in the stem at distances δx_u and δx_d, respectively from the upward and downward ends of the heated segment. They give the temperatures T_{u1} , T_{u2} ; T_{d1} and T_{d2} (Fig. 1). The lateral heat flux is estimated by measuring the temperature difference δT_{lat} between the two faces of a neoprene tape surrounding the heating jacket, with a copper-constantan thermopile.

The upstream and downstream conductive heat flows, q_u and q_d (W) are:

$$q_u = -K_u (T_{u1} - T_{u2}) = -K_u \, \delta T_u \qquad \text{Equation 4}$$

$$q_d = -K_d (T_{d2} - T_{d1}) = -K_d \, \delta T_d \qquad \text{Equation 5}$$

with: $K = ks / \delta x$, where k is the thermal conductivity of wood (0.4 to 0.6 W m^{-1} $^\circ$C^{-1}); s is the cross sectional area (m^2) of the trunk; δx is the distance (m) between the measuring points of the temperature differences δT_u and δT_d ($\delta x = 1$ cm for a tree with a 6-10 cm diameter).

The lateral heat flux, q_{lat} (W) is:

$$q_{lat} = -K_{lat} \, \delta T_{lat} \qquad \text{Equation 6}$$

with K_{lat} (W $^\circ$C^{-1}) computed *in situ* when the sap flow is zero.

The term of heat stored q_{sto} (W) in the volume delimited by the sensor, neglected by Sakuratani (1981), Baker & Van Bavel (1987), Steinberg, Van Bavel & McFarland (1989), can become very large at some times of the day, specially after sunrise (Fig. 2). It must be considered and is evaluated as:

$$q_{sto} = c_{wood} \, V \, dT/dt \qquad\qquad \text{Equation 7}$$

where c_{wood} is the heat capacity of trunk (2.6 to 3.0 10^6 J kg^{-1} $^{\circ}$C^{-1}); V is the volume (m^3) considered in the heat balance ($V = s \, (L + \delta_{xu} + \delta_{xd} + \delta x)$); dT/dt is the rate of variation of the mean temperature in the heated segment. We have shown that dT/dt can be suitably estimated from the variation of the temperature T_{cent} measured at the center of the heated segment with a fifth copper-constantan thermal probe (Nasr, 1989; Valancogne, Nasr & Angelocci, data to be published).

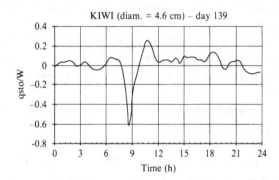

FIG. 2. Daily course of the relationship between the rate of heat (q_{sto}) stored in the heated stem segment and the heat power (W) produced by the heating jacket observed in a kiwi plant during a sunny day.

The temperature increase of the sap across the heated segment δT is:

$$\delta T = (T_{d1}+T_{d2}) / 2 - (T_{u1}+T_{u2}) / 2)$$
$$= \delta T_u / 2 + (T_{d1} - T_{u1}) + \delta T_d / 2) \qquad \text{Equation 8}$$

So, the sap flow can be computed from five measurements: δT_u, $(T_{d1}-T_{u1})$, δT_d, δT_{lat} and T_{cent}.

The conduction coefficient of the fluxmeter, K_{lat} (W $^{\circ}$C^{-1}), is:

$$K_{lat} = - \text{MINIMUM} \, [W - (q_d - q_u) - q_{sto}) / \delta T_{lat}] \qquad \text{Equation 9}$$

when the sap flow is near zero. These conditions are likely to occur only at the end of the night and after rains or irrigations. For some species of fruit trees (for instance apple trees), dew or smog can also be necessary.

This heat balance method was tested under greenhouse conditions on potted apple trees with different diameters (4 to 8 cm). The sap flow was compared to the transpiration of each tree measured by weighing them with an electronic balance. Afterwards, this method was used on apple trees in high density orchards (stem diameters: 4 to 10 cm), but without reference measurements of transpiration.

RESULTS AND DISCUSSION

The hourly computed sap flow closely follows, with a small delay, the measured transpiration (Fig. 3). These small discrepancies would be strongly increased if the heat storage q_{sto} was ignored (Fig. 3). This delay can be explained by a water contribution of the foliage and the stem to transpiration. Therefore, it is not possible strictly to evaluate the method on a hourly basis.

The daily computed sap flow (Fig. 4) remains within the limits of 10% around the daily measured transpiration.

The heat balance method can give some wrong results if not much care is taken at the time of the setting up. The electrical insulation of the thermocouples must be excellent. The heating jacket and the thermal fluxmeter must be firmly set up around the stem. A thermal insulation is added around the sap flow sensor to reduce the effects of rapid temperature variations in the environment; it must be uniform to avoid additional perturbations of the sap flow sensor.

The disposition of the thermal sensors must follow a few rules for the measurements to be appropriate for computing the sap flow.

If the sap flow is near zero, with a symmetric setting of upward and downward pairs of thermal sensors ($\delta x_u = \delta x_d$), δT is nearly zero and the use of equation 3 leads to very high computed sap flow even if the convective heat flow is very small. An asymmetrical configuration with $\delta x_u < \delta x_d$ avoids this problem, which could be met especially during the night.

A problem of δT being less than or equal to 0 can also arise if the direct solar radiation reaches the lower part of the stem: the ascending sap can then be much warmer than inside the heated segment. This situation occurs in orchards at midmorning and sometimes, but with smaller effects, at midafternoon. A suitable insulation of the lower part of the stem is indispensable.

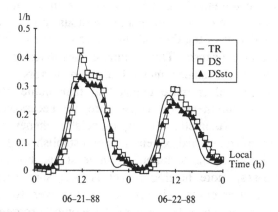

FIG. 3. Daily course of the computed sap flow (*DS*sto) and the measured transpiration (*TR*) for a potted apple tree (stem diameter: 6.3 cm) under greenhouse conditions. The sap flow computed without considering the heat storage (*DS*) is added to illustrate the importance of the heat storage.

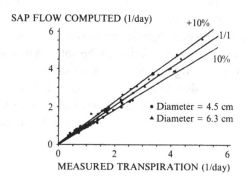

FIG. 4. Relation between computed daily sap flow and measured daily transpiration for two potted apple trees with different diameters.

δT can also be underestimated for large values of sap flow, because the heat does not get to the center of the stem within the limits of the heated segment. An increase of δx_d and setting up the thermocouples of the thermal sensors at a depth 1/4 to 1/3 of the stem radius prevents an overestimation of the sap flow in the high flow conditions.

This heat balance method has the advantage of measuring directly the sap flow in the whole stem of plants without any calibration: the conduction coefficient of the thermal fluxmeter K_{lat} is estimated after the setting of the sap flow sensor. This method causes much less disturbance than the former heat balance method used in sectors of tree trunk. However, this method is more demanding than some other thermal methods: five measurements of temperature are necessary. It is very useful for small trees (diameter <10 cm), especially whenever the position and the extent of functional xylem is not established. We have not examined the limits of the technique in large stems; however, as stem diameter increases, more heating power is necessary, and additional battery power may preclude field operation. Moreover, the heat storage (the estimation of which is less accurate) becomes progressively more significant in the heat balance.

The heat balance method is now used to estimate the daily water requirements of trees in high density orchard and to establish their relation with the climate and the characteristics of the orchard (crop coefficients).

Now, diverse thermal methods are available to scientists for measuring the sap flow in stems or trunks in different conditions. It is possible to study very precisely the course of drought in relation to other parameters (water potential, stomatal resistance, variations of the stem diameter, surface temperature of leaves). Special emphasis is put on:
- daily course of the sap flow in relation to the solar radiation;
- daily water use by well irrigated or water stressed trees.

These methods are also helpful for investigations concerning the distribution of water in plants. They should play a great part in the analysis of the different strategies of plant adaptation to drought conditions.

REFERENCES

Baker, J. M. & Van Bavel, C. H. M. (1987) Measurement of mass flow of water in the stems of herbaceous plants. *Plant, Cell and Environment*, **10**, 777-782.

Čermák, J., Kucera, J. & Penka, M. (1976) Improvment of the method of sap flow rate determination in full grown trees based on heat balance with direct electric heating of xylem. *Biologia Plantarum (Praha)*, **18** (2), 105-110.

Cohen, Y., Fuchs, M. & Green, G. C. (1981) Improvement of heat pulse method for determining sap flow in trees. *Plant Cell and Environment,* **4**, 391-397.

Daum, C. R. (1967) A method for determining water transport in trees. *Ecology,* **48** (3), 425-431.

Granier, A. (1985) Une nouvelle méthode pour la mesure du flux de sève brute dans le tronc des arbres. *Annale des Sciences Forestières,* **42**, 193-200.

Marshall, D. C. (1958) Measurement of sap flow in conifers by heat pulse transport. *Plant Physiology* (Bethesda), **33**, 385-396.

Nasr, Z. (1989) Mesure du flux de sève brute dans un arbre par une méthode thermique. Evaluation de la méthode et application. *Thèse de Doctorat d'Etat, option Ecologie Générale,* Université de Paris-Sud (France).

Sakuratani, T. (1981) A heat balance method for measuring sap flow in the stem of intact plant. *Agricultural Meteorology,* **37**, 9-17.

Steinberg, S., Van Bavel, C. H. M. & McFarland, M. J. (1989) A gauge to measure mass flow rate of sap in stems and trunks of woody plants. *Journal of the American Society of Horticultural Science,* **114**, (3), 466-472.

Swanson, R. H. & Whitfield, D. W. (1981) A numerical analysis of heat pulse velocity: theory and practice. *Journal of Experimental Botany,* **32** (126), 221-239.

Valancogne, C. & Nasr, Z. (1989a) Measuring sap flow in the stem of small trees by a heat balance method. *HortScience,* **24**, 383-385.

Valancogne, C. & Nasr, Z. (1989b) Une méthode de mesure du débit de sève brute dans de petits arbres par bilan de chaleur. *Agronomie,* **9**, 609-617.

Heat pulse measurements on beech (*Fagus sylvatica* L.) in relation to weather conditions

J. HEIMANN[1] AND W. STICKAN [2]

[1] *Institut für Forstbotanik, Universität Göttingen, Büsgenweg 2, D-3400 Göttingen, Germany.*
[2] *Systematisch-Geobotanisches Institut, Universität Göttingen, Untere Karspüle 2, D-3400 Göttingen, Germany.*

SUMMARY

The spatial and temporal variation of heat pulse velocity was investigated in a 150-year-old beech tree (*Fagus sylvatica* L.). High correlation was found between heat pulse velocity and water vapour saturation deficit of the air (84%) and photosynthetic photon flux density (together 85%). No additional influence of temperature and wind speed could be detected.

INTRODUCTION

The water budget of trees is linked to the water potential gradient between soil and atmosphere. Transpiration of a tree crown depends on stomatal conductivity, radiation, air temperature, water vapour saturation deficit of the air and wind speed. Variations are influenced by the soil water potential, the conducting tissues, the root system and the amount of leaves. The base of the trunk is the optimal place to measure transpiration of a whole tree assuming that storage is negligible.

The intention of this research work was to complement a programme of gas exchange measurements by heat pulse velocity measurements. The spatial and temporal variation of heat pulse velocity was investigated in a single tree. The correlation between heat pulse velocity and weather conditions during the vegetation period was analysed.

MATERIALS AND METHODS

The determinations were carried out on a single beech tree growing in a beech stand (*Luzulo-Fagetum*) which has been intensively investigated during the last 20 years (B1 research area of the IBP, International

Biological Programme, "Solling-Projekt", Germany; Ellenberg, Mayer & Schauermann, 1986).

In addition to gas exchange measurements (Schulte *et al.*, 1989; Stickan *et al.*, 1991) heat pulse velocity was recorded as a relative measure for transpiration or water uptake of a whole tree (beech no. B68, which has already been studied by Schulze, 1970).

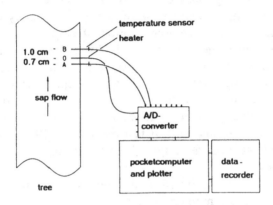

FIG.1. Heat pulse meter with installed measuring system.

The heat pulse method used was developed by Huber (1932) and modified by Swanson (1962, 1967). A portable heat pulse meter was constructed at the Systematisch-Geobotanisches Institut of the University of Göttingen (Fig.1). A pocket computer (PC 1500, Sharp) controlled the measurements and recorded the data. Heat pulse velocity (v, cm s^{-1}) was calculated according to Closs (1958):

$$v = (OB-OA) (2\,t_0)^{-1} \qquad \text{Equation 1}$$

where: t_0 = time from the beginning of heating until the temperatures in A and B have equal values; OA, OB = distance between heat source and temperature sensor A or B.

Temperature sensors (0.15 cm in diameter) were installed 1.0 cm deep in the wood 0.7 cm below and 1.0 cm above the heater (soldering iron).

Diurnal and annual courses of heat pulse velocity were registered in relation to air temperature, relative air humidity, photosynthetic photon flux density (PPFD) and wind speed.

RESULTS AND DISCUSSION

The diurnal course of heat pulse velocity followed the courses of water vapour saturation deficit of the air and photosynthetic active radiation without an obvious lag-phase caused by the water storing function of the tissues (Fig. 2). The diurnal course of heat pulse velocity corresponded well with the course of the transpiration of single leaves. Heat pulse velocity was higher in branches and changed faster with changing weather conditions compared to the base of the trunk (Fig. 2). In branches heat pulse velocity was more closely coupled with atmospheric conditions. This suggests that trunk tissues have a water storing capacity.

Heat pulse velocity reached values up to 0.2 m h^{-1}. These values seem to be very low compared with data from Huber (1935). Huber used a different method and he might have overestimated the velocities (Marshall, 1958).

Heat pulse velocity differed at different sides of the trunk. It was always higher at the northern and eastern side and here it could be measured earlier in the morning and later in the evening than the other two sides. These differences could be due to the asymmetrical shape of the crown and root system.

On rainy days no movement of the xylem sap could be registered. Sudden rain events reduced the velocity abruptly..

When water vapour saturation deficit of the air was high, sap flow did not cease at night due to the high cuticular transpiration of beech leaves.

Sap flow started immediately after bud break in May and stopped in the middle of September. By comparing similar days of different seasons the heat pulse velocity was found to be positively dependent on the amount of foliage.

The annual course of heat pulse velocity was recorded from July to September 1987 and from May to July 1988. Daily mean values of heat pulse velocity during the vegetation period were correlated with mean values of water vapour saturation deficit of the air (Fig. 3), with radiation (photosynthetic photon flux density, PPFD) (Fig. 4), air temperature (Fig. 5) and wind speed (Fig. 6).

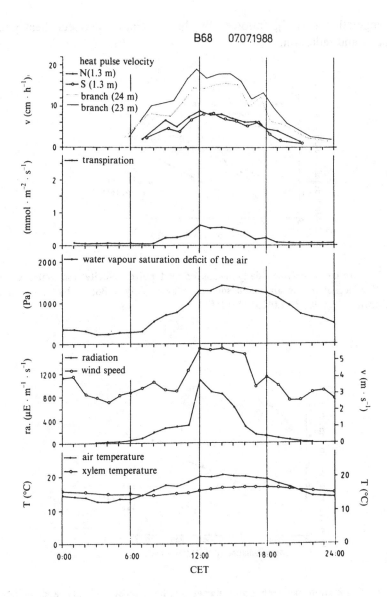

FIG. 2. Heat pulse velocity was measured at two points (1.3 m) at the base of the trunk of beech no. B68 and at two branches (23 m, 24 m). Transpiration of single leaves and radiation (photosynthetic photon flux density) were measured in a cuvette in the light crown.

The highest correlation was found between heat pulse velocity and water saturation deficit (bud break time could be differentiated from the rest of the vegetation period), followed by the correlation between heat pulse velocity and radiation.

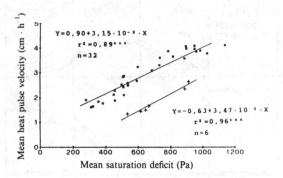

FIG. 3. Correlation between daily means of heat pulse velocity and water vapour saturation deficit of the air. Solid squares: May to July 1988; +: bud break time. Significance: * 5.0%; ** 1.0%; *** 0.1%.

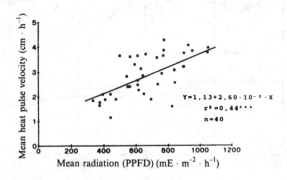

FIG. 4. Correlation between daily means of heat pulse velocity and radiation (photosynthetic photon flux density) from May to July 1988.

For this a multiple correlation ($r^2 = 0.85$, $n = 48$ diurnal courses) was calculated for the whole measuring period:

$$y = 1.057 + 2.6 \cdot 10^{-3} x_1 + 4.7 \cdot 10^{-4} x_2 \qquad \text{Equation 2}$$

where y = heat pulse velocity (cm h^{-1}), x_1 = saturation deficit (Pa) and x_2 = PPFD radiation (mmol photons m^{-2} h^{-1}).

Heat pulse velocity was highly correlated with water vapour saturation deficit (84%) and with radiation (photosynthetic photon flux density) (together 85%). Little influence of air temperature and wind speed was found although air temperature is related to the water vapour saturation deficit.

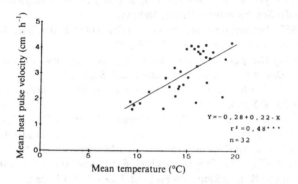

FIG. 5. Correlation between daily means of heat pulse velocity and air temperature from May to July 1988.

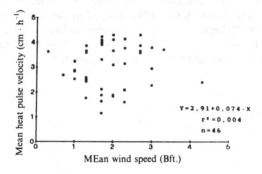

FIG. 6.. Correlation between daily means of heat pulse velocity and wind speed from May to July 1988.

The correlation was linear as the water supply in the Solling mountains is sufficient and does not limit transpiration of the trees. Thus no influence of the soil water potential on heat pulse velocity could be detected as only once was a value lower than -30 kPa determined during the measuring period.

REFERENCES

Closs, R.L. (1958) The heat pulse method for measuring rate of sap flow in a plant stem. *New Zealand Journal of Science*, **1**, 281-288.

Ellenberg, H., Mayer, R. & Schauermann, J. (eds.) (1986) *Ökosystemforschung - Ergebnisse des Sollingprojekts*. Ulmer, Stuttgart.

Huber, B. (1932) Beobachtung und Messung pflanzlicher Saftströme. *Berichte der Deutschen Botanischen Gesellschaft*, **50**, 89-109.

Huber, B. (1935) Die physiologische Bedeutung der Ring- und Zerstreutporigkeit. *Berichte der Deutschen Botanischen Gesellschaft*, **53**, 711-719.

Marshall, D.C. (1958) Measurement of sap flow in conifers by heat transport. *Plant Physiology*, **33**, 385-396.

Schulte, M., Wahle, E., Stickan, W. & Runge, M. (1989) Der CO_2- und H_2O-Gaswechsel von Buchen im Solling als Indikator für latente Schadstoffwirkungen. *Berichte der Deutschen Forschungszentrums Waldökosysteme* (Reihe A), **49**, 67-78.

Schulze, E.-D. (1970) Der CO_2-Gaswechsel der Buche (*Fagus sylvatica* L.) in Abhängigkeit von Klimafaktoren im Freiland. *Flora*, **159**, 177-232.

Stickan, W., Schulte, M., Kakubari, Y., Niederstadt, F., Schenk, J. & Runge, M. (1991) Ökophysiologische und biometrische Untersuchungen in einem Buchenbestand (*Fagus sylvatica* L.) des Sollings als ein Beitrag zur Waldschadensforschung. *Berichte der Deutschen Forschungszentrums Waldökosysteme* (Reihe B), **18**, 1-82.

Swanson, R.H. (1962) *An instrument for detecting sap movement in woody plants*. U.S.D.A. For. Serv. Rocky Mountain For. Range Exp. Sta. Pap. No. 68, 16 pp.

Swanson, R.H. (1967) Improving tree transpiration estimated based on heat pulse velocity measurements. In: 14. IUFRO-Kongreß, München, Bd. **1**, 252-263, DVFFA Deutscher Verband Forstlicher Forschungsanstalten.

Extremely fast changes of xylem water flow rate in tall trees caused by atmospheric, soil and mechanic factors

J. ČERMÁK AND J. KUCERA

*Institute of Forest Ecology, University of Agriculturae
Brno, Czechoslovakia.*

SUMMARY

Xylem water flow rate as estimated with the stem tissue heat balance (THB) method in mature trees is a very sensitive indicator of different phenomena which can occur at any part of the soil-plant-atmosphere continuum. Some situations were observed when the conductive xylem was subjected to cavitation processes under natural and experimental conditions in coniferous and broadleaved species. The response of xylem water flow rate at the stem base to rain and rarely occurring abrupt depletion of soil water was compared with its response when tree trunks or branches were excised. The experiment represents the fastest possible change in flow that could ever occur. The response of flow in the trunk was slower when branches or tree tops of spruce were removed. The fastest changes observed under natural conditions for all species were also slower. The response of flow to rain was only half that caused by removal of trunks in birch, oak and maple. The abrupt depletion of soil water moisture was followed by a change in flow rate still less than in the case of rain for oak growing in sandy soil. The process of rapid soil water depletion typical of sandy soil is usually much slower in clay or loamy soils. Nevertheless, cavitation in xylem vessels can be expected anywhere and the records of xylem water flow rate made with sufficiently high resolution show a rapid decrease under high evaporation rate, which can be easily detected with TBH method. Some examples are given which suggest cavitation even under moderate water stress. The results obtained can be used to assess the extent to which the conducting system is endangered by drought and consequently to predict the tree survival.

INTRODUCTION

Devices for measuring vertical water transport in xylem, enabling detection of even very fast changes in the water flow rate, have been developed recently (Čermák, Deml & Penka, 1973; Čermák *et al.*, 1982; Kucera, Čermák & Penka, 1977). When studying the xylem water flow in tall trees, we have focussed on the fast flow rate changes, trying to relate the causes of these changes to the speed with which they take place, *i.e.*

acceleration and/or deceleration. The aim of the study was to explore the technical limits of the flow measuring device. We also tried to find out how rapid changes in the xylem water flow rate take place in tall trees within physiological limits, when caused by factors influencing both leaves and roots.

MATERIALS AND METHODS

The xylem water flow rate was studied in tall trees growing under different experimental conditions (Table 1).

TABLE 1. Characteristics of trees sampled for the various experiments.

Species	DBH (cm)	H (m)	MP (m)	Soil Type	Experiment
Acer platanoides L.	10-12	10-12	1.3	IL	stem cut-off
Betula alba L.	15	18	1.3	IL	rain
Quercus petraea Matt. (Liebl.)	10	11	1.3	IL	stem cut-off
Quercus petraea	42	22	3.0	SD	drought
Picea abies (L.) Karst.	25	25	1.3	BL	branch cut-off
Fagus sylvatica L.	50	26	4.0	CL	irrigation

IL - illimerized soil with admixtures of loess
SD - sandsoil
BL - brown loamy soil
CL - clay soil , DBH - diameter at breast height, H - height of tree, MP - measuring point.

The xylem water flow rate (Q_w) was measured by the tree-trunk heat balance (THB) method with internal (direct electric) heating of plant tissues (Čermák *et al.*,1973; Čermák *et al.*,1982; Kucera *et al.*, 1977). The Q_w was calculated from the applied heat power (P) and the temperature difference (dT) at the measuring point:

$$Q_w = a\,(P\,/\,dT) - b \qquad\qquad \text{Equation 1}$$

where a is a constant (including physical and spatial characteristics) and b represents the heat losses from the measuring point, caused by other reasons than by the xylem water flow. The values of b were recorded as the "fictitious flow" (Q_w^{fic}), at the time when the true flow was approaching zero (the predawn value) and must be subtracted from the recorded values of flow (Q_w^{rec}) to get the true flow ($Q_w = Q_w^{rec} - Q_w^{fic}$). A battery of five electrodes was applied with a set of four thermocouples compensating for the influence of changes in ambient temperature (Čermák & Kucera, 1981).

Short periods of measurement were analyzed in this study, where the flow rate changed rapidly after the experimental trees were subjected to the influence of specific natural factors such as sudden rain showers or abrupt soil water depletion. Experimentally, several trees were cut off close to flow measuring points (maples and oaks - see Table 1) and in spruce a substantial fraction (1/2) of the branches from the middle and lower part of the crown and later the top (2 m long) were cut off quickly (branches within one minute) during the measurement. Experiments took place under conditions of sufficient moisture (with the exception of large oak and beech trees) and fine weather in the summer.

RESULTS AND DISCUSSION

Removal of the stem close to the measurement point provides an opportunity to study the fastest possible rate of change in water flow. Such a situation, as recorded with the THB method working with constant temperature difference (constant dT) and/or constant heat power (constant P), is shown in Fig. 1.

The response of the measuring system with constant P is monotonous, without oscillations and with a maximum rate of change of about 30% per minute. The response of the measuring system with constant dT on the same change in flow was found to be faster (as it should be in principle), being as much as 60% per minute. This response, in fact, is the fastest response of the regulative system to the change of input variables in the measuring device applied. In the extreme case during the decrease of flow to zero, the response to changing flow rate is accompanied by an oscillation, suggesting short-term negative flow. This oscillation can be eliminated partially by technical means, *i.e.* by optimizing the central part of the electronic regulator, but certain deformation of the response to the actual stoppage of the flow with the measuring system is unavoidable. The principial reason for this is the inertia of the heat field within the measuring point after such a rapid change. This phenomenon can be used practically just to indicate the flow stoppage. The negative flow that we observed from the point of excision downwards to the roots when the excision is made above the measuring point (see Čermák *et al.*, 1984) at least partially corresponds to the reality, as observed following rapid exposure of xylem lumen to the atmospheric pressure. However, the response was similar even in the case when the cut was made just below the measuring point. In general, the response of flow in mature trees corresponds to that observed in shoots (Rychnovska, Čermák & Smid 1980; Čermák *et al.*,1984). In any case, the magnitude of the change measured with the constant dT system, corresponds more closely to reality than with the use of constant P system, and can be interpreted as the maximum detectable response of the xylem water flow in large trees we can measure in field conditions.

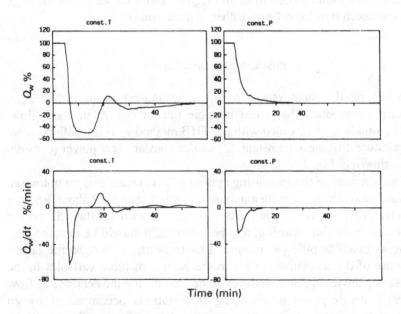

FIG. 1. The course of xylem water flow (Q_w) and its acceleration (Q_{wt}/dt) as observed during short time intervals (about one hour) during a fine summer day in a mature maple tree *(Acer platanoides* L.) after the tree-trunk was cut off just above the measuring point. A measuring device with constant power (P) or constant temperature difference (dT) was applied experimentally. Maximal rate of change in flow (deceleration) was about -30% and -60% min^{-1}.

A sudden decline of xylem water flow rate caused by stoppage of transpiration due to a rain shower can occur in natural conditions. This situation is illustrated on the chart record using the measuring system with constant dT in birch during a fine summer day in which afternoon showers occurred (Fig. 2).

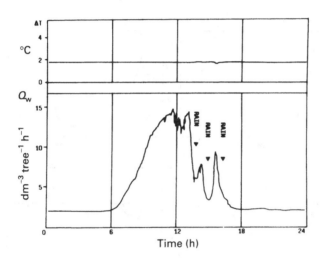

FIG. 2. Diurnal course of xylem water flow rate (Q_w) in a mature birch tree (*Betula alba* L.) under non-limiting soil moisture conditions on a warm summer day with showers in the afternoon (a copy of the chart record). Q_w was measured with the TBH method using a constant $dT = 1.7\ ^{\circ}C$, as shown in the upper part of the figure. The scale is shifted in the figure, *i.e.* true Q_w zero line is at the value of 2 dm^3 tree^{-1} h^{-1}, below that is the fictitious flow only.

From the values of flow rate recorded, "fictitious flow" equal to 2 dm^3 tree^{-1} h^{-1} (as shown at night) should be subtracted. dT remained constant over the entire period. The decrease in flow rate caused by rain was practically as fast as its increase when fine weather resumed (this is striking especially after the second shower). A detailed analysis of the record from this and one other day (Fig. 3) showed that the fastest changes in the xylem water flow rate (when transpiration stopped due to a severe rainstorm) were about 20% per minute. This is safely below the detection limits as found in experiments with the cut off stems; also the absence of oscillations confirms that the response as recorded is reliable. The measurement of the same change in flow with the constant P system

186   J. ČERMÁK AND J. KUCERA

could distort the true response (it could simulate a slow response only) even when quantitative data taken for the longer time intervals (say hours) would be correct.

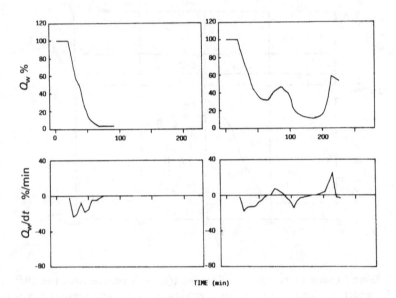

FIG. 3. The course of xylem water flow (Q_w) and its acceleration (Q_w/dt) as observed within short time intervals (about three hours) during fine summer days with some afternoon showers in a mature birch (*Betula alba* L.). The line at 100% Q_w corresponds to the simulation just before the rain. The influence of one and of two subsequent showers is shown on the left and right part of the figure. Maximal rate of change in flow (deceleretion) was about -20% min⁻¹.

A marked response of xylem water flow rate was also observed in experiments with a mature spruce tree when one half of branches were cut-off quickly from middle and lower parts of the crown and the next experimental day the tree-top was cut off (Fig. 4).

The first treatments represent the removal of 1/3 and the second 1/4 of the total needle dry mass. The experiment confirmed that the specific daily transpiration of needles in the upper crown was about three times higher in comparison with that in middle and lower crown as analyzed in another study (Čermák & Kucera, 1990). The speed of changes in xylem water flow rate at the stem base (about 20 m below the crown centre) reached about 5% per minute and was thus lower than in the birch.

In contrast, we observed more rapid changes in xylem water flow rate in experiments with a smaller oak, whose crown was cut off 5 m above the measuring point. Certain time-lag was also observed between the accurately defined moment of the cut and the onset of the response. The oak started to respond after 30 seconds, the change in xylem water flow rate thus moved at 20 cm s^{-1}. This value should be taken as minimal only, due to the inertia of the measuring system and recorder.

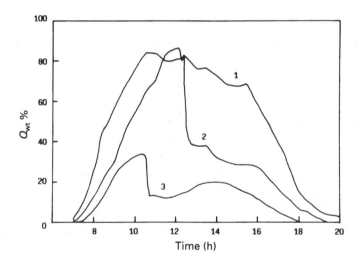

FIG. 4. Diurnal courses of xylem water flow rate (Q_{wt}) in a mature spruce tree (*Picea abies* (L.) Karst.) under non-limiting soil moisture on three warm summer days. No experiments were made in the first day (1), but 1/2 of the branches from the middle and lower crown were cut off during the second experimental day (2) and the tree-top (2 m long) was cut off during the third experimental day (3).

We were fortunate during the long-term measurement of xylem water flow rate in another mature oak growing on sandy soil under increasing drought conditions, to record a brief period (or a phenomenon appearing repeatedly during subsequent days) when flow decreased dramatically (Čermák *et al.*, 1981). The record of such a response during two days of measurement (Fig. 5) shows that the change in flow reached about 10-15% per minute.

In this case the record is considered faithful. Nevertheless, the slope of the decrease, caused possibly by interruption of contacts between soil capillary water and the root surface, perhaps combined with the cavitation in xylem vessels, is very remarkable. (On the fourth day the flow reached about 1/5 of the original values only.) We also found in another study (Čermák, Kucera & Matyssek, in press) that after long-term severe drought the xylem water flow rate of a big beech tree immediately responded to heavy irrigation (150 mm per 20 min). The speed of change in flow rate, being more pronounced immediately after the onset of irrigation and gradually decreasing with saturation of soil by water, reached about 4% per minute.

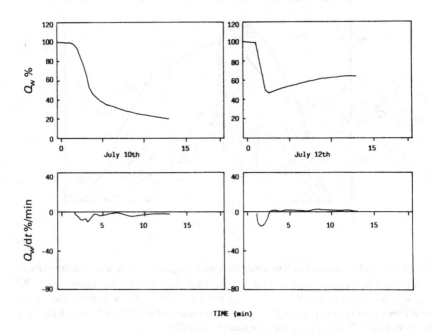

FIG. 5. The course of xylem water flow (Q_w) and its acceleration (Q_w/dt) as observed within short time intervals (about 20 minutes) during two fine summer days of measurement in a mature oak tree (*Quercus petraea* (Matt.) Liebl.) growing in sandy soil. The line at 100% Q_{wt} corresponds to the situation just before the abrupt soil water depletion. Maximal rate of change in flow (deceleration) was about -10% min[-1].

Generally, we can conclude that the measuring device applied is capable of detecting even very fast changes in xylem water flow rate, which appear under natural conditions in coniferous and broadleaved tall trees. Changes in flow are similarly rapid under fast changes of environmental factors influencing both leaves (crown) and roots. This corresponds to the idea of xylem water continuum in the conductive elements of the tree stems (see the review *e.g.* by Zimmermann, 1983). Therefore, the speed of flow changes was found to be similar in species such as oak and spruce with very different diameters of tracheal elements, and also with flow velocities differing by an order of magnitude (see Hinckley, Lassoie & Running, 1978). The water reserves in tree trunks cannot be used under such rapid changes in the xylem flow rate, evidently due to high hydraulic resistances between water reservoirs and the conductive elements.

REFERENCES

Čermák, J., Deml, M. & Penka, M. (1973) A new method of sap flow rate determination in trees. *Biologia Plantarum* (Praha), **15**, 171-178.

Čermák, J., Jenik, J., Kucera, J. & Zidek, V. (1984) Xylem water flow in a crack willow tree (*Salix fragilis* L.) in relation to diurnal changes of environment. *Oecologia* (Berlin), **64**, 145-151.

Čermák, J. & Kucera, J. (1981) The compensation of natural temperature gradient in the measuring point during the sap flow rate determination in trees. *Biologia Plantarum* (Praha), **23**, 469-471.

Čermák, J. & Kucera, J. (1990) Changes in transpiration of healthy mature trees due to environmental conditions and of those with damaged water conductive system. In *Proc.Cs.MAB Natl.Co./IUFRO Internat.workshop "Verification of hypotheses and possibilites of recovery of forest ecosystems"* (ed. E. Klimo & J. Materna), pp. 275-286. Beskydy Mountains, Czechoslovakia, Sept.4-8, 1989. Univ. Agr. Brno.

Čermák, J., Kucera, J.& Matyssek, R. (1991) Primary causes of tall beech decline in *Fagetum* "typicum on heavy soils" (in Czech). *Lesnictrí* (in press).

Čermák, J., Kucera, J., Stepànkova, M., Prax, A. & Kuráz, V. (1981) The transpiration flow in tall trees, soil hydrolimits and the course of soil water content at the housing-estate of Hodonin/Bazantnice (In Czech). *Research Report of the Institute of Forest Ecology, Agricultural University,* Brno, 60 pp.

Čermák, J., Ulehla, J., Kucera, J. & Penka, M. (1982) Sap flow rate and transpiration dynamics in the full-grown oak (*Quercus robur* L.) in floodplain forest exposed to seasonal floods as related to potential evapotranspiration and tree dimensions. *Biologia Plantarum* (Praha), **24**, 446-460.

Hinckley, T.M., Lassoie, J.P. & Running, S.W. (1978) Temporal and spatial variation in the water status of forest trees. *Forest Science Monograph,* **20**.

Kucera, J., Čermák, J. & Penka, M. (1977) Improved thermal method of continual recording the transpiration flow rate dynamics. *Biologia Plantarum* (Praha), **19**, 413-420.

Rychnovská, M., Čermák, J. & Smid, P. (1980) Water output in a stand of *Phragmites communis* Trin. A comparison of the three methods. *Acta Scientiarum Naturalium* (Brno), **14**, 1-27.

Zimmermann, M.H. (1983) *Xylem Structure and the Ascent of Sap.* Springer-Verlag, Berlin.

Water relations and water transport
in coppice vs. single stem *Quercus cerris* L. trees

M. SABATTI, G. E. SCARASCIA MUGNOZZA,
R. VALENTINI AND A. DEL LUNGO

Dipartimento Scienze dell'Ambiente Forestale e delle sue Risorse,
Università della Tuscia, via S. Camillo de Lellis, 01100 Viterbo, Italy.

SUMMARY

The coppice and coppice-with-standards treatments are still widely applied in broadleaf forest stands, especially in Southern Europe. These sylvicultural systems are also promising for the management of fast growing tree crops and for multilayer agroforest stands in tropical regions.

Coppice sprouts show a higher productivity on a per leaf area basis compared to trees grown from seed and allowed to develop naturally as 'standards'. Evidence is presented that coppice trees had also improved water relations and greater water transport on a per leaf area basis.

INTRODUCTION

Forests are naturally regenerated either via sexual reproduction or by vegetative propagation. In the first case, the production of seeds, their germination and the establishment of seedlings can be encouraged by several silvicultural techniques.

On the other hand, regeneration of many species can occur by growth and development of the meristematic centres located at the stump or root levels. The sprouts and suckers thereby produced give rise to a forest stand called coppice; sometimes, within coppice stands some trees are left uncut for a maximum of three to four consequent harvests, to provide some regeneration from seeds, in order to replace the declining stumps. These trees left for sexual reproduction are called standards. Coppice and coppice-with-standards are old silvicultural systems, widely used for firewood and timber production in Southern Europe. They have received renewed attention in recent years because they represent a valid option for new silvicultural goals such as fast growing biomass plantations (Ferm & Kauppi, 1990) and tropical agroforestry systems (Stewart, 1980).

A vast amount of scientific information is available concerning coppice growth and biomass production but there is little work on the physiological basis of the higher growth rates of coppice compared to standards and to high stand (Auclair, 1986). To this aim, the phenomenon of juvenility is generally invoked, but it would be appropriate to analyze the different physiological determinants of yield (Blake & Raitanen, 1981). The objective of this note is to examine water transport of sprouts versus standards, within a particular oak coppice stand.

MATERIALS AND METHODS

The research was conducted in a 1300 ha forest, located west of Viterbo, in Central Italy (42° 23' N, 11° 55' E), on a relatively high plateau at an elevation of about 160 m. Bedrock belongs to volcanic-sedimentary rocks developed during the Quaternary period; the soil can be classified as a Acquic Haploxeralf, of brown colour, characterized by poorly differentiated soil layers. It has a reduced content of organic matter, a subacid reaction and a high content of clay. The vegetation is typical of the Supra-Mediterranean zone and is made up of deciduous tree species. The dominant species is *Quercus cerris* L. associated with *Q. petraea* Liebl. in some parts of the forest. In the tree layer other species such as *Carpinus orientalis* Mill., *Acer campestre* L., *Fraxinus ornus* L., *Ulmus minor* Mill. and *Pyrus* spp. can also be sporadically found.

The forest is managed as a coppice with standards, with 2400 sprouts on about 1200 stumps per ha and with about 100 standards per ha. The rotation interval is about 18 years for the coppice and a multiple of it (*i.e.* three or four times) for the standards. The mean annual aboveground woody biomass increment is 7 tons of dry matter ha^{-1} yr^{-1} (Corona, La Marca & Schirone, 1986). Within a section of the forest, of 15 years of age, a 100x100 m^2 experimental plot was located and fenced. All the trees within the plot were mapped and their height and d.b.h. measured. Basal area was 23 m^2 ha^{-1} for coppice and 5 m^2 ha^{-1} for standards. A scaffolding enabled access to crowns of two standard trees (25 cm in diameter at breast height (d.b.h.) and 18 m high) and two coppice sprouts (d.b.h. 12

cm, height 15 m) representing trees of different sizes and different positions within the canopy.

Sap flux of the four study trees was measured by means of a heat pulse velocity meter developed in our Department, according to Swanson (1962). However, only one temperature sensor, above the heating element, was adopted as proposed by Miller, Vavrina & Christensen (1980) and the heat pulse velocity was calculated taking the time at first onset. The instrument included a double digital chronometer, a timer and a heating element, 2 mm in diameter, made up of a nickel wire coiled around a thin needle and covered by a teflon sleeve. Temperature was measured by a type K thermocouple whose signal was amplified by an AD amplifier and read by a voltmeter. Sap velocity measurements were carried out in the stem at about 1.3 m from the ground; the thermocouple was inserted into the stem 1.8 cm above the heating element, within the outer conducting ring. In order to eliminate one cause of variation, the probes were implanted in the south side of the stem (Miller *et al.*, 1980). Once holes were made, they were filled with glycerol to prevent air from coming in and to improve thermal conduction. After inserting the two probes, thermal insulation was applied and the entire assembly was covered with aluminum foil.

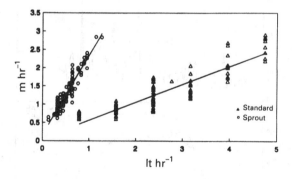

Fig. 1. Relationship between water uptake of a standard and a sprout tree and their apparent sap velocity.

The thermal probe was heated for 2 s until a temperature of 45 °C +/-1 °C was reached. Calibration of the instrument was required in order to derive from the apparent sap velocity (HPV) the transport sap velocity (TV) and, therefore, the sap flux density. For calibration, two trees were

M. SABATTI *et al.*

chosen: a coppice sprout (12 cm d.b.h.) and a standard tree (25 cm d.b.h.). First, early in the morning a plastic tank was arranged around the stem of the calibration tree. All the fissures were sealed with silicone rubber to avoid leakage; then, the tank was filled with water. In order to allow the tree to take up the water from the tank, the stem was girdled, below the water level, to a depth of 3 cm. A meter was inserted in the tank to measure the change of water level and a plastic film was laid down to avoid evaporation from the tank. The HPV meter was then installed and sap velocity together with the water level in the tank were recorded every 5 to 10 min.

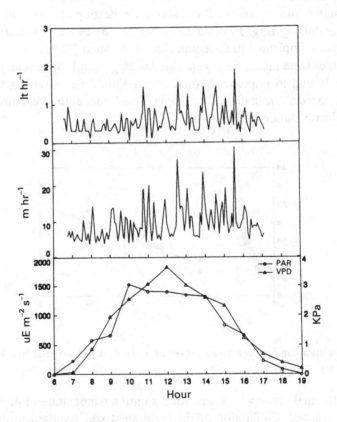

FIG. 2. Water uptake, transport sap velocity and environmental parameters on a sunny day for a sprout tree.

At the end of the calibration day a dye was dissolved in the water left in the tank in order to identify the xylem vessels that were conducting the sap. It should be recognized that water in the tank was at atmospheric pressure and therefore could deviate from normal condition in the field. However, our intention was to compare the heat pulse velocity measurement with water uptake by the tree, under the same conditions.

Later, the two calibration trees were felled; then cross sections of the stem were cut above and below girdling, brought to the lab and observed at the stereoscope to measure the characteristics of the xylem. The felled trees were also utilized for leaf area and other morphometric measures, as will be described later.

Sap flux velocity of the four study trees was measured throughout two consecutive years, 1986 and 1987, from the months of June through October, for a total of 7 days per year. Over the course of each day, measurements started at 6:00 h to 8:00 h solar time and lasted to the evening; every 10 to 15 min sap velocity was recorded for each tree. Data were then processed in the laboratory by means of a personal computer.

Leaf conductance to water vapour was measured according to a stratified sampling scheme. The crown of the coppice sprouts was ideally divided into two horizontal layers, the lower one from 9 to 12 m and the upper one from 12 to 15 m, whereas in the crown of the standards a third layer was also identified, from 15 m on. Within each layer 4 to 6 leaves were sampled on each study tree, taking care that sunny and shady aspects of tree crowns were equally represented in the measurements. Leaf conductance was measured with a steady state porometer (Beardsell, Jarvis & Davidson, 1972) Li-Cor 1600 (Li-Cor, Lincoln, NE, USA) whose calibration was checked every 2-3 days; each entire cycle of measurements took 2 hours. The values of stomatal conductance were averaged over an entire measuring day to give the mean daily stomatal conductance.

Predawn leaf water potential was also frequently measured throughout the field experiments.

Meteorological parameters were regularly monitored by means of a standard agrometeorological field station of the National Research Council (CNR), installed in an open field, approximately 1 km away from the study plot. The station automatically recorded air and soil temperature, air relative humidity, wind speed and direction, and photosynthetic active radiation (C.A.E., Bologna, Italy).

During the middle of the season, in 1987, 12 trees, chosen to represent the different size classes of the study plot and including the trees used for

the HPV meter calibration, were utilized for measures of crown
architecture and leaf area vertical distribution (Scarascia Mugnozza,
unpublished data).

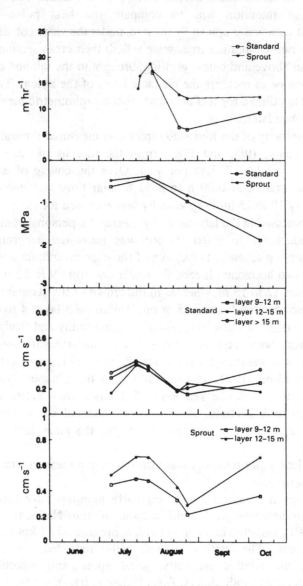

FIG. 3. Mean daily transport sap velocity, base water potential and mean daily stomatal
conductance in standard and sprout trees in 1986.

RESULTS

Calibration procedures demonstrated a good relationship between water uptake by trees and the apparent sap velocity as measured by the HPV meter (Fig. 1).

Even though a relatively large scattering of the data was observed, the correlations gave high R^2 values, such as 0.9 and 0.8, respectively, for coppice sprout and the standard. The two slopes are, as expected, quite different because at a given sap velocity there was a larger volume of water flowing through the larger conducting area of the standard tree.

In Table 1 the characteristics of the xylem cross section of the sprout and standard tree are reported. The use of a dye, however, showed that only the vessels of the outer ring were functional. Normalizing the values of water uptake by the area of the conducting cross-section of the xylem (see Miller *et al.,* 1980) gives the transport sap velocity ; the regression equations between the apparent and the transport velocities are the following:

$$TV_{st} = 7.56015 \ HPV_{st} + 1.968100 \qquad \text{Equation 1}$$

and

$$TV_{sp} = 6.81382 \ HPV_{sp} - 0.162325 \qquad \text{Equation 2}$$

respectively, for standard and sprout.

TABLE 1. Dimension of the trees and their xylem (VD_{mean} = mean vessel diameter , VD_{max} = maximum vessel diameter, VA = cross section surface of the vessels of the last ring).

	height (m)	d.b.h. (cm)	VD_{mean} (μm)	VD_{max} (μm)	VA (mm^2)
Standard	18	25	285	400	212.5
Sprout	15	12	280	340	59.5

198 M. SABATTI *et al.*

In Fig. 2 the trends of the water uptake and of transport velocity of the sprout tree, during the calibration day, are reported. Values of sap transport velocity varied between a minimum of 4 m hr^{-1} to a maximum of 30 m hr^{-1}.

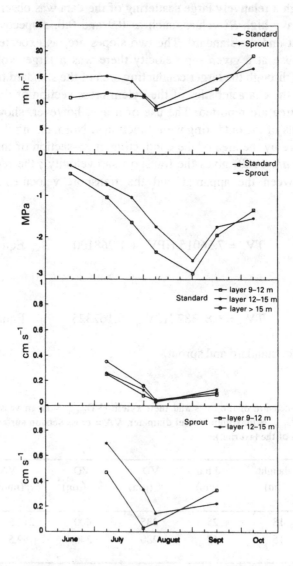

FIG. 4. Mean daily transport sap velocity, base water potential and mean daily stomatal conductance in standard and sprout trees in 1987.

Fluctuations of the velocity values throughout the day were quite apparent although there was a clear increase of the sap velocity toward the afternoon.Furthermore, fluctuations of transport sap velocity corresponded to fluctuations of water uptake.

Sap transport velocity, throughout the two experimental growing seasons, had a trend that corresponded to the seasonal variations of the most important physiological and environmental parameters. During the year 1986, characterized by a slightly dry summer season, the mean daily transport velocity remained quite high, around 12 to 19 m hr $^{-1}$ for the sprout and at lower values, 6 to 17 m hr $^{-1}$, for the standard (Fig. 3). In fact, predawn water potential of the study trees remained at relatively high values in that year and reached the maximal values (-0.4 MPa) in July, in correspondence to the highest values of transport velocity. At the same time, stomatal conductance also showed high values for both sprouts (0.7 cm s^{-1}) and standards (0.4 cm s^{-1}) during the months of July and August. In 1987 (Fig. 4) climate was quite different from the previous year and caused the progressive decrease of predawn water potential to minimum values such as -2.5 MPa for sprout and -2.8 MPa for standard.

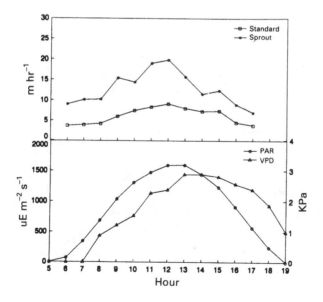

FIG. 5. Vapour pressure deficit (VPD), photosynthetically active radiation (PAR) and transport sap velocity of standard and sprout trees, on a sunny day (August 22, 1986).

In the second half of September the occurrence of abundant rainfall allowed the improvement of the plant water potential which rose to -1.3 and -1.5 MPa. Also the daily transport velocity correspondingly increased from 9 m hr^{-1} at midsummer to higher values in September and October (12 to 19 m hr^{-1}). Again, in the year 1987 a higher transport velocity was measured in coppice sprouts compared to standards although the differences were not as evident as during the previous year and, at the end of the season, sap velocity was slightly higher in the standards.

The differences in mean velocity among pairs were tested for significance according to a two-sample t test; three out of four and three out of seven were significant for 1986 and 1987, respectively. The same test applied to predawn water potential and stomatal conductance means showed significant differences for both years.

In order to examine the trend of transport velocity throughout the measuring days of the two study years, mean hourly values were first computed. The daily trend of such values was, as expected, strongly associated with daily changes of environmental factors including solar radiation and vapour pressure deficits.

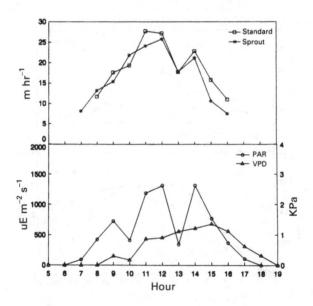

FIG. 6. Vapour pressure deficit (VPD), photosynthetically active radiation (PAR) and transport sap velocity of standard and sprout trees, on a cloudy day (October 9, 1987).

On clear sunny days (Fig. 5) the curve of sap transport velocity was rather bell shaped and followed the variation of the vapour pressure deficit (VPD) and of the photosynthetic active radiation (PAR). The maximum was quite accentuated for the sprout tree (20 m hr $^{-1}$) and rather low for the standard tree (9 m hr $^{-1}$). During cloudy days (Fig. 6), the pattern of sap transport velocity was clearly associated to the changing PAR. Both, sprout and standard trees showed high values of transport velocity with a maximum of 28 m hr $^{-1}$ and 26 m hr $^{-1}$ for standards and sprouts, respectively.

The greater transport velocity values presented in Fig. 6 compared to those of Fig. 5 can be explained by the improved water relations existing in the month of October 1987 compared to August 1986. In fact, in the period July-August 1986 rainfall was less than 50 mm whereas in the period September-October 1987 it reached 300 mm.

The relationship between sap transport velocity and the environmental variables such as vapour pressure deficit and solar radiation was tested by regression analysis; together, the two variables could explain about 73% of the entire variation of sap velocity in sprout trees and 41% in standards. Vapour pressure deficit was more important than solar radiation. The regression of the hourly values of plant water potential on sap transport velocity showed high values of the determination coefficient, 0.6 and 0.8 respectively, for sprout and standard trees. Apparently, for a given increment of sap transport velocity the plant water potential decreased faster in standards.

DISCUSSION

The heat-pulse velocity technique is generally recognized as valuable for relative studies rather than for the absolute determination of sap flow rates in trees (Jones, Hamer & Higgs, 1988). The velocity of a heat pulse does not always coincide with that of the sap and may underestimate it by as much as four times (Marshall, 1958; Leyton, 1967) in conifers and diffuse-porous hardwoods. In the present experiment, the heat pulse velocity was underestimated by a factor of about seven, possibly because of the high flows observed in ring-porous *Quercus cerris* trees (see also Jones *et al.*, 1988). Other reasons for the observed discrepancies may be found in the phenomenon of embolism of vessels observed in the sample trees during the summer period and in the disturbance to the conducting system caused by the insertion of heating and temperature measuring devices into the wood. However, the absolute calibration of the sensor output in terms of total sap flow through the stem, as performed in this experiment, showed a good relationship between heat-pulse velocity technique and water uptake values (Figs. 1 and 2). We hypothesize that the fluctuations of sap flow observed in *Quercus cerris* L., during calibration, are determined, principally, by rapid variations of micrometeorological parameters at the canopy level. These fluctuations are transmitted relatively rapidly down along the stem in ring-porous species such as the *Quercus* and may affect the velocity of the sap stream. Absolute calibration of this technique also gave good results in *Eucalyptus marginata* Sm. as shown by Doley & Grieve (1966). Sap transport velocities displayed large variation throughout the measuring days, in the present experiment; however, maximum velocity was 30 m hr^{-1} in accordance with the observations of Huber (1956) in *Q. robur* L. and Miller *et al.* (1980) in *Q. alba* Lam. and *Q. velutina* L.

TABLE 2. Mean sap flux (SF), sap flux/leaf area ratio (SF/LA), xylem cross surface per unit of foliage area (VA/LA) and Huber value in standard and sprout trees.

	SF (l hr^{-1})	SF/LA (l m^{-2} hr^{-1})	VA/LA (mm^2 m^{-2})	Huber value (mm^2 g^{-1})
Standard	2.47	0.016	1.4	0.56
Sprout	0.87	0.065	4.5	1.3

The efficiency of aboveground biomass production per unit of leaf area was twice as large for coppice sprouts (151 g m^{-2}) compared to standards (70 g m^{-2}) in the highest fertility plots of the study forest; this difference decreased with stand fertility so that leaf efficiency became 54 g m^{-2} and 40 g m^{-2}, for sprouts and standards respectively in the lowest fertility class plots. Causes of these relatively large differences in leaf efficiency between the two types of trees could in part reside in the different ages of sprouts, 15 years old, compared to standards, which were on average 50 years old. However, some interesting physiological variations were also found in water relations. Measurements of sap velocity showed significantly higher values in sprouts compared to standards, throughout the two study years. Upon transformation of sap velocity into sap flux, sprout trees showed a mean sap flux of 0.9 l hr^{-1} over the two years, whereas standards had 2.5 l hr^{-1}. On a unit of leaf area basis, however, coppice sprouts had values 4 times greater than standards, 0.065 l m^{-2} hr^{-1} vs. 0.016 l m^{-2} hr^{-1} (Table 1). Possible explanations for this elevated water flow per unit of leaf area reside, partly, in the higher value of the ratio of conducting vessel area to leaf area characteristic of sprouts compared to standard trees (Table 2).

This translates into an improved water status, *i.e.* higher water potential, shown by the coppice trees and into a greater average leaf conductance and therefore higher transpiration rates. It can be concluded that coppice sprouts behave more as water spender, compared to standards. Another interesting consequence of the improved water relations of the coppice is that it will, probably, be less subject to symptoms of tree decline associated with water stress (Vannini, 1987).

REFERENCES

Auclair, D. (1986) Coppice versus single stem: physiology, growth, economics. *Proceedings XVIII IUFRO World Congress, Ljubljana, Yugoslavia*, Division I, Vol. II, p. 759.

Beardsell, M. F., Jarvis, P. G. & Davidson, R. (1972) A null balance diffusion porometer suitable for use with leaves of many shapes. *Journal of Applied Ecology*, **9**, 677-690.

Blake, T. J. & Raitanen, W. E. (1981) *A summary of factors influencing coppicing.* International Energy Agency, Rep. NE **22**, 1-24.

Corona, P., La Marca, O. & Schirone B. (1986) Ricerche sull'ottimizzazione dell'intensità della matricinatura nei cedui di cerro: il ceduo composto a maturità. *Annali Accademia Italiana di Scienze Forestali*, **XXXV**, 3-38.

Doley, D. & Grieve, B. J. (1966) Measurement of sap flow in a eucalypt by thermo-electric methods. *Australian Forest Research*, **2**, 3-27.

Ferm, A. & Kauppi, A. (1990) Coppicing as a means for increasing hardwood biomass production. In *Forestry, forest biomass, and biomass conversion: the IEA bioenergy agreement (1986-1989) Summary Reports* (ed. C.P. Mitchell, L. Zsuffa, S. Anderson & D. J. Stevens), pp. 107-121. Elsevier Applied Science, London and New York.

Huber, B. (1956) Gefassleitung. In *Handbuch der Pflanzenphysiologie* (ed. E. W. Ruhland), **33**, 541-582. Springer, Berlin.

Jones, H. G., Hamer, P. J. C. & Higgs, K. H. (1988) Evaluation of various heat-pulse methods for estimations of sap flow in orchard trees: comparison with micrometeorological estimates of evaporation. *Trees*, **2**, 250-260.

Leyton, L. (1967) Continuous recording of sap flow rates in tree stems. In *IUFRO-Kongress*, München., Paper 1, Vol. I, Sect.01-02-11, pp. 240-249.

Marshall, D. C. (1958) Measurement of sap flow in conifers by heat transport. *Plant Physiology*, **33**, 385-396.

Miller, D. R., Vavrina, C. A. & Christensen, T. W. (1980) Measurement of sap flow and transpiration in ring porous oaks using a heat pulse velocity technique. *Forest Science*, **3**, 485-494.

Stewart, P. J. (1980) Coppice with standards: a system for the future. *The Commonwealth Forestry Review*, **59**, 149-154.

Swanson, R. H. (1962) *An instrument for detecting sap movement in woody plants*. U.S.D.A. Forest Service. Rocky Mt. Forest and Range Experimental Station Paper 68.

Vannini, A. (1987) Osservazioni preliminari sul deperimento del cerro (*Quercus cerris* L.) nell'Alto Lazio. *Informatore fitopatologico*, **37**, 54-59.

Environmental control of water flux
through Maritime pine (*Pinus pinaster* Ait.)

D. LOUSTAU[1] AND A. GRANIER[2]

[1] *I.N.R.A. Station de Recherches Forestiers,*
Laboratoire d'Ecophysiologie et Nutrition,
BP. 45 Gazinet, 33610 Cestas, France.
[2] *I.N.R.A. Station de Sylviculture et Production,*
Laboratoire d'Ecophysiologie et de Bioclimatologie,
BP 35 Champenoux, 54280 Seichamps, France.

SUMMARY

This paper outlines the effects of environmental factors on the liquid and gaseous water flux through mature Maritime pine in a stand from southwest France. The sap flow density, expressed per unit area of the cross-section of sapwood, was measured throughout two growing seasons in the bole of a sample of trees. Stomatal conductance (g_s), needle water potential, water table depth and soil water content were measured periodically. From daily courses of needle water potential and sap flow density, the bulk soil-leaf specific conductance (L_s) was computed. The soil water deficit revealed the most important environmental factor controlling the sap flow. It reduced the sap flow density by decreasing both the gradient of water potential between the soil and the leaves, and the bulk hydraulic conductance along this circuit. The reduction of bulk specific conductance might be attributed to variations in soil water potential for the day. In a similar way, a multifactorial model demonstrated the close dependence of g_s on the predawn water potential and air vapour saturation deficit. This behaviour supports the hypothesis of stomatal control of cavitation.

INTRODUCTION

Maritime pine stands extend over more than four million hectares in southern Europe. In the natural area of this species, which includes the western Mediterranean coasts, and Atlantic coasts of Portugal, Spain and France up to the Loire estuary, summer drought lasts from one to three months. Even so, environmental and physiological control of water consumption and transpiration of Maritime pine (*Pinus pinaster* Ait.) are still poorly documented (Gash *et al.*, 1989; Granier *et al.*, 1990; Diawara, Loustau & Berbigier, 1991).

Water flow through a tree may be partitioned into a liquid phase and a gaseous phase. Environmental regulation of water flow in a tree may also be described according to these two phases.

In the liquid phase, sap flow between the soil and the leaves of a plant, J (kg h^{-1}), can be described by the general transport equation. It depends on a driving force, the difference in water potential between the leaves (Ψ_l) and the soil-root interface (Ψ_{sr}), and on the bulk hydraulic conductance between them, L.

$$J = L \, (\Psi_l - \Psi_{sr}) \qquad\qquad \text{Equation 1}$$

Environmental regulation of sap flow is exerted through these two parameters. Coniferous species behave as "drought avoiders" and keep their minimal leaf water potential above a specific threshold. Therefore, the difference of water potential across the tree fluctuates between a minimal value imposed by the gravitational force, in conditions of zero-flux, and maximal values typically less than 1.5 MPa. This limitation sets an upper limit to the sap flow through the tree. In the course of a drought, the drop in soil water potential reduces the difference in water potential between soil and leaves and lowers this upper limit of the sap flow.

A decrease of bulk hydraulic conductance has also been shown to occur under drying soil conditions for different species of trees (Running, 1980; Cohen, Moreshet & Fuchs, 1987; Peña & Grace, 1986; Granier et al., 1989; Reich & Hinckley, 1989). This decrease induces a supplementary reduction in sap flow.

Gaseous diffusion of water from a coniferous tree crown, E, depends on the atmospheric evaporative demand, which is defined by the water vapour pressure difference between leaf and air ($e_l - e_a$), and on leaf surface conductance to vapour diffusion, g_w.

$$E = g_w \, (e_l - e_a) \qquad\qquad \text{Equation 2}$$

For coniferous species, the transpiration rate, E, is closely controlled by the canopy conductance g_c, which is the main component of g_w (Jarvis & McNaughton, 1985). The response of canopy conductance to environmental variables is well documented for several coniferous species (Tan & Black, 1976; Lindroth, 1985; Stewart, 1988) but still not for Maritime pine, apart from the previous results of Gash et al. (1989) on wet soil.

Results of a study concerning the environmental regulation of water flow through mature Maritime pine under field conditions are presented here. Conclusions are based on previously published data (Loustau, Granier & El Hadj Moussa, 1990) and new information, focusing mainly on variations in hydraulic and gaseous conductances regulated by environnemental conditions, is presented.

MATERIALS AND METHODS

Experimental site

The experimental site has already been described by Diawara *et al.,* (1990). The Bray Forest is situated 20 km southwest from Bordeaux (44° 42' N. 0° 46' W.), in the northern part of the "Landes de Gascogne" Forest. Mean annual temperature and total precipitations were 12.5 °C and 932 mm respectively. The soil was a sandy humic podzol with a cemented Bh horizon limiting the depth of the root zone at 75 cm. A permanent water table fluctuated between 20 and 150 cm under the soil surface.

The stand of pines, originating from a sowing in 1970, covered an area of 16 ha and its present density is 880 stems per hectare. The mean height in 1989 was 12.5 m, the mean diameter at breast height (dbh) was 0.18 m and the leaf area index of pines was estimated to be around 3.

Micrometeorological variables (air temperature, water vapour saturation deficit, wind speed, precipitation and global irradiance) were measured at 18 m high, from a tower situated in the middle of the stand. From these measurements Penman's evapotranspiration (PET) was computed.

Soil water content was measured every ten days with a neutron scattering probe, using a set of ten access tubes covering an area of 900 m², including the sampled trees.

Sap flow measurements and hydraulic conductance estimation

The continuous heating method, extensively described by Granier (1987), was used to measure the sap flow density, expressed per unit area of sapwood. Sapwood area of each tree was estimated from three cores extracted with an increment borer, heartwood and sapwood being separated according to their transparency in diffuse light. This method had previously been calibrated against direct measurements (Diawara *et al.*, 1991). In 1988, measurements of sap flow extended from March to November. In 1989, they started in May and ended at the end of October. Mean sap flow density for the whole stand, J_s (kg dm^{-2} d^{-1}), was estimated by the arithmetic mean of sap flow densities of the sample of ten trees (Loustau *et al.*, 1990). The total cross-sectional area of sapwood of the stand (A_s, dm^2 m^{-2}) was estimated once a year from a sampling of 33 trees and daily transpiration, E_p (kg m^{-2} d^{-1}, or mm d^{-1}), of the stand was computed as follows:

$$E_p = A_s J_s \qquad \text{Equation 3}$$

Needle water potential, Ψ_l (MPa), was measured using the pressure chamber technique on a sample of eight pairs of one-year-old needles taken from the south side of the three upper whorls of the same tree (13 m high, 0.20 m dbh).

The bulk hydraulic specific conductance, L_s (kg h^{-1} MPa^{-1} dm^{-2}), and the water potential at zero flux, Ψ_o (MPa), were estimated by linear regression of hourly values of sap flow density, J_s (kg m^{-2} h^{-1}), against water potential (Cohen *et al.*, 1987) and on the same tree, according to the following equation:

$$L_s = J_s / (\Psi_l - \Psi_o) \qquad \text{Equation 4}$$

Estimation of surface conductance

Following Lindroth (1985), the specific surface conductance, g_s (cm s^{-1}), was defined as surface conductance per unit leaf area.

Samples of 20 to 30 pairs of needles from the same tree, stratified according to whorl height, azimuth and needle age class, were measured for ten days in 1988, from April to November. The sample was reduced to eight pairs of needles of the same tree in 1989 and was measured for only two days, in June and September. Measurements were made with a null-balance automatic porometer (Li-Cor 1600, Li-Cor Inc, Lincoln NE, USA) on the same tree measured for water potential. As only slight variations of leaf conductance were found according to shoot position within the crown (Loustau & El Hadj Moussa, 1989), hourly values of g_s were estimated from porometric measurements by the arithmetic mean of the measured values.

A multifactorial non-linear regression of hourly values of g_s against air vapour saturation deficit, d_e (Pa), predawn water potential, P_b (MPa), and total irradiance, S_g (W m^{-2}), was computed. Preliminary analysis showed air temperature did not significantly contribute to the variance of g_s. Formulae from Stewart (1988) and Halldin *et al.* (1980) were used:

$$g_s = g_{sm} [1-a (d_e)] \qquad \text{Equation 5}$$

$$g_s = g_{sm} [S_g / (S_g+b)] \qquad \text{Equation 6}$$

$$g_s = g_{sm} [1-c (d-P_b)] \qquad \text{Equation 7}$$

where g_{sm} is the maximal measured conductance (0.5 cm s^{-1}) and a,b,c and d are parameters fitted by the model. The multifactorial model was a geometric combination of these equations:

$$g_s = g_{sm} (1-a (d_e)) S_g / (S_g+b) (1-c(d-P_b)) \qquad \text{Equation 8}$$

RESULTS

Seasonal course of daily sap flow and transpiration of pines

The precipitation for spring and early summer 1988 exceeded long-term averages, while 1989 spring and summer precipitation was uncommonly low. The seasonal evolution of the ratio between PET and E_p (Fig. 1) showed that a reduction in the transpiration of pines occurred at the beginning of September in 1988 and at the end of June in 1989. In both years, this reduction occurred when the water content of soil-root zone dropped below 55 mm (REWC=23 %). Above this threshold, the ratio of E_p to PET remained close to 0.6 and transpiration of trees fluctuated within 3 and 4 mm for clear days. Below a REWC of 23 %, daily transpiration fluctuated principally with the occurrence of precipitation events which temporarily restored soil water availability. Nevertheless, it did not exceed 2 mm d^{-1}, with a minimum value of 0.15 mm d^{-1}.

Water relations

For the predrought period, the surface conductance dropped from 0.3 cm s⁻¹ in the morning to 0.1 cm s⁻¹ in late afternoon. For the drought, the specific surface conductance was reduced 3 to 5 fold, and did not significantly change all through the day. Sap flow density was reduced by the same magnitude (data not shown).

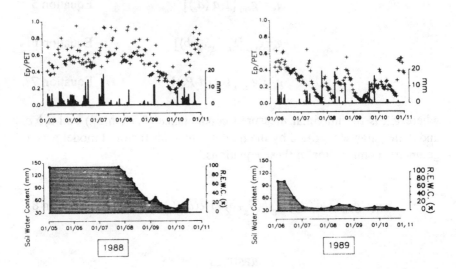

FIG. 1. (above) Seasonal evolution of the ratio between transpiration (E_p) and Penman's evapotranspiration (PET) (+), and rainfall; (below) soil water content, expressed in absolute values (mm) and as a percentage of available water (REWC, %).

Minimum daily values of water potential were largely independent of soil water content and fluctuated between -1.4 and -1.8 MPa (Fig. 2).

Predawn water potential dropped sharply when REWC was below 23%. The lowest values for predawn water potential was 1.2 MPa but lower values could have occurred for the driest periods.

Variations in sap flow

As a consequence of the drop in predawn water potential, the soil-leaf water potential difference was greatly reduced when REWC dropped under 23%. This reduction accounted for one half of the total depletion of sap flow density. Reduction of absolute values of the slope of regression lines between sap flow density and water potential (Fig. 3) demonstrated the reduction of bulk hydraulic conductance when the soil extractable water content dropped below 23 % (Table 1).

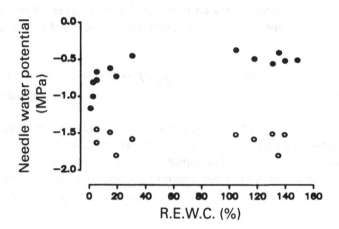

FIG. 2. Relation between predawn (full circles) and midday (open circles) values of needle water potential and relative water content of soil.

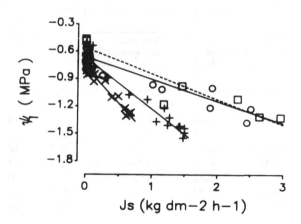

FIG. 3. Hourly values of sap flow density related to corresponding needle water potential for four selected days with REWC of, respectively, 100 (open squares), 80 (open circles), 15 (+) and 5 % (x).

Computed values of the zero-flux water potential (Table 1) were not significantly different than predawn water potential.

Surface conductance modelling

The multifactorial model gives an r^2 of 0.8 for 66 degrees of freedom. Computed values for the different parameters lead to the following relation:

$$g_s=0.5(1-0.0004\,(d_e))\,(S_g\,/\,(S_g+93))\,(1-1.52\,(-0.24-P_b))$$

Equation 6

Seventy five per cent of the total variance is explained by the predawn water potential and vapour saturation deficit, while 5% may be attributed to the radiation component. The minor contribution of irradiance could be attributed to the lack of measurements of g_s early in the morning and late in the evening, when values of photosynthetically active radiation were low.

TABLE 1. Relative extractable water content of soil (REWC), predawn water potential (Ψ_b), and computed values of zero-flux potential and soil to leaf conductance for seven selected days.

REWC (%)	Predawn water potential (MPa)	Zero flux potential (MPa)	Bulk specific conductance (kg dm^{-2} h^{-1} MPa^{-1})	r^2	n
		Computed parameters of the regression between water potential and sap flow density			
150	-0.56	-0.66	4.16	0.92	15
140	-0.41	-0.41	2.97	0.86	14
110	-0.49	-0.50	3.43	0.95	14
22	-0.45	-0.56	3.58	0.91	14
15	-0.62	-0.63	1.72	0.94	20
08	-0.78	-0.84	1.26	0.98	11
05	-0.68	-0.74	1.21	0.93	19

Discrepancy between predicted and observed values of g_s for the fitted model may be explained by changes in stomatal efficiency in the course of the growing season (Lindroth, 1985) and errors in micrometeorological measurements. Figures 4, 5 and 6 show some results of the model.

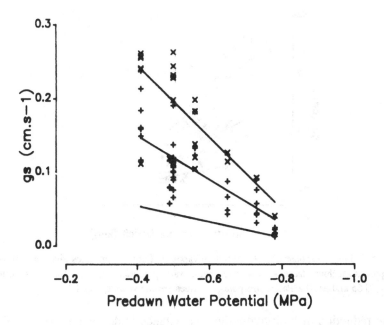

FIG. 4. Response of g_s to the predawn needle water potential; straight lines give the simulated responses for different classes of saturation vapour deficit (from the top to the bottom, 500 and 1250 Pa). Corresponding observed values are shown as x and +.

DISCUSSION

The pattern of water potential of Maritime pine that develops over the course of the growing season is similar to that found for other species of the genus *Pinus, e.g. Pinus contorta* (Running, 1980) or *Pinus radiata* (Benecke, 1980) under natural conditions. The onset of the decrease in predawn water potential begins at 23% REWC. This value is quite close to that found for Douglas fir (Running, Waring & Rydell, 1975) but higher values were reported (Granier, 1987). Differences might be attributed to the difference in site characteristics and to the different methods used in estimating the REWC.

It is somewhat difficult to compare our values of L_p with others, because of the computational methods and related units used. Values of hydraulic conductance of Maritime pine are in the upper range of values for coniferous species as compared with *Pinus contorta* (Running, 1980),

Picea abies or *Abies bornmulleriana* (Granier *et al.,* 1989).

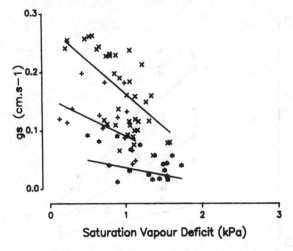

FIG. 5 (below). Response of gs to saturation vapour deficit; straight lines give the simulated responses for three classes of predawn needle water potential (from the top to the bottom: -0.4, -0.65 and -0.80 MPa). Corresponding values are shown as x, + and *.

The reduction of bulk hydraulic conductance under water stress in field conditions has already been shown for coniferous (Running, 1980; Granier *et al.,* 1989) and hardwood species (Cohen *et al.,* 1987; Reich & Hinckley, 1989). Compared with these results, hydraulic conductance of Maritime pine seems to be very sensitive to predawn water potential (Table 1). The changes in hydraulic conductance during soil dessication may include both a soil-root component (i) and a root to leaf component (ii).

(i) On dry soil, low values of soil hydraulic conductance could induce a reversible depletion of water potential in the vicinity of the root during the day, when roots are absorbing water. In this experiment, hourly fluctuations of soil water potential could not be measured. However, the rate of drainage of the water table below the root zone was higher from 11:00 to 18:00 (IST) than during the night. This indicated a higher rate of capillary rise from the water table, which implied the occurrence of a temporary decrease in water potential in the root zone.

(ii) Variations in hydraulic conductance within the trees could also have occurred. Questions arise about the location of this loss of conductance and the nature of the mechanisms involved. The lowest measured values of water potential seem much too high to induce cavitation and embolism in the stem or in the thicker branches of the tree, if compared with values given by Tyree & Sperry (1988) for four woody species.

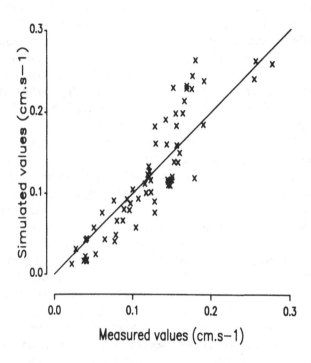

FIG. 6. Computed versus observed values of g_s. Straight line is the bisector.

However, xylem vulnerability to cavitation for this species is still unknown. Peña & Grace (1986) have shown that cavitation in the stem of Scots pine might explain a loss of hydraulic conductance under water stress. Therefore, further investigations on the soil-root interface and the hydraulic architecture of Maritime pine would be necessary in order to understand the variations in hydraulic conductance through the soil-tree system.

The transpiration of Maritime pine of "les Landes" region also exhibited a high sensitivity to soil water deficit. It should be noticed that reduction in transpiration under soil drying conditions (Fig. 1) does not take into account a possible decrease of sapwod area throughout the season, since A_s was assumed to be constant. Such a decrease could induce a supplementary reduction of transpiration. However, this is presumably only a minor component of the variation of the stand transpiration, since year to year variation of sapwood area did not exceed 8%.

The fitted model of g_s only explains 80% of the variance and residuals are not randomly distributed (Fig. 6). So, a better relation might perhaps be fitted to these data. At the present stage predawn water potential is the main variable contributing to the canopy specific conductance. This relation had already been stated by Loustau *et al.* (1990). Strong correlation between predawn water potential and surface conductance has been found for other tree species (Reich & Hinckley, 1989). However, the sensitivity of g_s to predawn water potential seems higher for Maritime pine than for most coniferous species. Similarly, in our experimental conditions, g_s was highly sensitive to air vapour saturation deficit. For example, an increase of the latter from 0.4 to 1.250 kPa would reduce g_s by 52%; such a reduction lies in the upper range of those reported by Sandford & Jarvis (1986) for different coniferous species.

The liquid and gaseous conductances of Maritime pine concurrently decreased when the predawn water potential went below -0.5 MPa. This behaviour could express an adaptative strategy to water stress and supports the hypothesis of stomatal control of cavitation (see Jones & Sutherland, in this volume). A high sensitivity of hydraulic conductance to soil-plant water status makes close regulation of transpiration necessary, especially for coniferous species. The turn-over of the conductive system of coniferous trees is slow and the regeneration of the entire system would extend over several years. Hence, cavitation and embolism must be prevented and water stress in conductive tissue must be avoided.

The drought avoidance behaviour exhibited by Maritime pine suggests that gaseous exchanges, dry matter production and growth of stands from southwest France would be closely controlled by the water balance of the site.

ACKNOWLEDGEMENTS

The authors thank M. Sartore for technical assistance, F. El Hadj Moussa and A. Ducasse for their participation in the experiment and M. Guedon for supplementary assistance. This experiment was supported by the INRA program "Bioclimatology of Maritime Pine Stands".

REFERENCES

Benecke, U. (1980) Photosynthesis and transpiration of *Pinus radiata* D. Don under natural conditions in a forest stand. *Oecologia*, **44**, 192-198.

Cohen, Y., Moreshet, S. & Fuchs, M. (1987) Changes in hydraulic conductance of Citrus trees following a reduction in wetted soil volume. *Plant Cell & Environment*, **10**, 53-57.

Diawara, A., Loustau, D. & Berbigier, P. (1991) Comparison of two methods for estimating the evaporation of a *Pinus pinaster* (Ait.) stand: sap flow and energy balance with sensible heat measurements by an eddy covariance method. *Agricutural & Forest Meteorology*, **55**, 49-56.

Gash, J.H.C., Shuttleworth, W.J., Lloyd, C.R., Andre J.C., Goutorbe J.P .& Gelpe, J. (1989) Micrometeorological measurements in Les Landes forest during Hapex-Mobilhy. *Agricultural and Forest Meteorology*, **46**, 131-147.

Granier, A. (1987) Evaluation of transpiration in a Douglas fir stand by means of sap-flow measurements. *Tree Physiology*, **3**, 309-320.

Granier, A., Bobay, V., Gash, J.H.C., Gelpe, J., Saugier, B. & Shuttleworth, W.J. (1990) Vapour flux density and transpiration rate comparisons in a stand of Maritime Pine *(Pinus pinaster* Ait.) in les Landes Forest. *Agricultural & Forest Meteorology*, **51**, 309-319.

Granier, A., Breda, N., Claustres, J.P. & Colin, F. (1989) Variation of hydraulic conductance of some adult conifers in natural conditions. *Annales des Sciences Forestiers*, **46 suppl.**, 357-360.

Halldin, S., Grip, H., Jansson, P.E. & Lindgren, A. (1980) Micrometeorology and Hydrology of Pine Forest Ecosystems - II Theory and Models. In *Structure and function of Northern Coniferous Forests. An Ecosystem Study* (ed. T. Persson). *Ecological Bulletin* (Stockholm), **32**, 463-503.

Jarvis, P.G. & McNaughton, K.G. (1985) Stomatal control of transpiration: scaling up from leaf to region. *Advances in Ecological Research*, **15**, 1-49.

Lindroth, A. (1985) Canopy conductance of coniferous forests related to climate. *Water Resource Research*, **21**, 297-304

Loustau, D., Granier, A. & Moussa El Hadj, F. (1990) Evolution saisonniére du flux de seve dans un peuplement de Pin maritime *(Pinus pinaster* Ait). *Annales des Sciences Forestiers*, **48**, 599-618.

Loustau, D. & Moussa El Hadj, F. (1989) Variability of stomatal conductance in the crown of a maritime pine *(Pinus pinaster* Ait). *Annales des Sciences Forestiers*, **46 suppl.**, 426-428.

Peña, J. & Grace, J. (1986) Water relations and ultrasound emissions of *Pinus sylvestris* L., before during and after a period of water stress. *New Phytologist*, **103**, 515-524.

Reich, P.B. & Hinckley, T.M. (1989) Influence of pre-dawn water potential and soil to leaf hydraulic conductance on maximum daily leaf diffusion conductance in two oak species. *Functional Ecology*, **3**, 719-726.

Running, S.W. (1980) Environmental and physiological control of water flux through *Pinus contorta*. *Canadian Journal of Forest Research*, **10**, 82-91.

Running, S.W., Waring, R.H. & Rydell, R.A. (1975) Physiological control of water flux in conifers: a computer simulation model. *Oecologia* (Berlin), **18**, 1-16.

Sandford, A.P. & Jarvis, P.G. (1986) Stomatal response to humidity in selected conifers. *Tree Physiology*, **2**, 89-103.

Stewart, J.B. (1988) Modelling surface conductance of Pine forest. *Agricultural & Forest Meteorology*, **43**, 19-35.

Tan, C.S. & Black, T.A. (1976) Factors affecting the canopy resistance of a Douglas fir forest. *Boundary-Layer Meteorology*, **10**, 475-488.

Tyree, M.T. & Sperry, J.S. (1988) Do woody plants operate near the point of catastrophic xylem dysfunction caused by dynamic water stress? *Plant Physiology*, **88**, 574-580.

Evaluation of transpiration of apple trees and measurement of daily course of water flow within the main branches of walnut trees

C. VALANCOGNE,[1] T. AMEGLIO,[2]
L. ANGELOCCI[3] AND P.CRUIZIAT [2]

[1] INRA, Centre de Bordeaux,
BP 81, 33883 Villenave d'Ornon CEDEX, France.
[2] INRA, Domaine de Crouelle,
63039 Clermont-Ferrand CEDEX, France.
[3] ESA Luiz de Queiroz, USP, Caixa Postal 9,
13400 PIRACICABA, SP, Brasil.

SUMMARY

The sap flow rate in trees is estimated from the heat balance computed in a segment of the trunk surrounded by a heating tape delivering a known rate of heating. Two pairs of thermocouples, radially inserted at the ends of the heated segment of the trunk, and a thermopile surrounding the heating tape are used to determine the different conductive heat flows and the heating of the sap. The rate of heat storage must be considered; it is computed from the measurement of the temperature in the heated volume.

This method was used to estimate the water use of apple trees in an orchard. The sap flow in the different trees increases with the leaf area. The mean sap flow rate of the trees is compared to the net radiation and to the radiation absorbed by the trees. The daily transpiration of the trees is lower than Penman potential evapotranspiration. The correlation with orchard net radiation or absorbed radiation is quite good.

The same method was used to compare sap flow of two walnut trees grown in containers and subject to normal and drought conditions. Sap fluxes were measured on the main trunk and on the three major branches supporting the foliage in order to test: a) the accuracy of the method; b) the relative importance of water flow within each branch.

Results show that: a) the rate of sap flow is very rapidly affected by the water stress conditions, more than the predawn water potential; b) in our case, due to the form and orientation of foliage, the changes in sap flow between the different branches, during a day, were synchronous. The computed sap flow within the trunk remains close to the sum of the fluxes of the three major branches; the observed discrepancies would be cancelled with the most recent improvements of the sap flow sensor.

INTRODUCTION

Most thermal methods give only the sap flux density (equivalent to a velocity). We have developed a heat balance method adapted from Sakuratani (1981) to measure directly the sap flow rate in small trees. In the first part of this paper, we will review the principle of the method. Then, we will give some examples of applications: first to the evaluation of water requirements of the fruit trees in orchards, and then to the measurement of the daily course of water flow within different branches of trees in different drought conditions.

Principles of the heat balance method

This method consists of measuring the different components of the heat balance of a stem segment which is heated by a heating jacket applied around the stem. These different components are:
- the downstream and upstream conductive heat flows,
- the lateral heat flow,
- an additional convective component due to the sap flow.

We have also to take into account the rate of heat storage changes with time. Each of these different terms is measured and the sap flow is deduced from the convective term (Valancogne & Nasr, 1989a, 1989b; Valancogne, Nasr & Angelocci, data to be published).

Estimation of the water use of apple trees in an orchard

The objective is to elucidate the relationship between the water requirements, the climate and the characteristics of the orchard (shape of the crown, dimensions, spacing and orientation of the rows).

A first experiment was made in 1988 in a six-year-old commercial orchard with Granny Smith apple trees, in the SW of France, with 4 m between the rows oriented N-S and 1 m between the plants along the rows. Water was continuously supplied by drip irrigation. Stem perimeters of the trees were measured in the orchard and sap flow was computed each 20 min from July to October in six representative trees instrumented with heat balance sensors. Leaf area measurements in the middle of August justified the choice of stem circumference as a criterion for selection of the trees.

The transpiration rate of the whole trees on an area of orchard basis (TR, mm h^{-1}) was estimated from individual measurements of sap flow and from distribution of the different sizes of the trees:

$$TR = N \, \Sigma \, (f_i \, d_{si})$$ Equation 1

where N is the number of trees per square meter of orchard; f_i is the frequency of the class i, d_{si} is the mean sap flow (l h^{-1}) of trees in the class i.

The daily transpiration (mm day^{-1}) was calculated by integration over 24 h. We have compared this transpiration with the PENMAN potential evapotranspiration given by meteorological stations (Fig. 1). Indeed in France, the PENMAN PET is often used for irrigation scheduling in orchards. The transpiration is lower than PENMAN PET except for the low value. This discrepancy can be explained by the difference of radiation interception pattern between usual crops and a row crop such as an orchard.

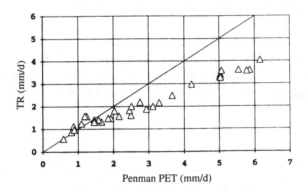

FIG. 1. Relation between daily transpiration (TR, mm d^{-1}) of the trees in an orchard and Penman Potential EvapoTranspiration (PET, mm d^{-1}) computed with net radiation above the orchard.

Transpiration of the apple trees was compared with absorbed global radiation (Fig. 2). Global radiation absorbed by the trees was calculated with a simple model of interception of the solar radiation by a row crop adapted from a vineyard model (Riou, Valancogne & Pieri, 1989). This model uses global and diffuse radiations, sun position computed from the date and time, geometrical characteristics of the tree rows, a "porosity coefficient" of the foliage in a row and reflection coefficients of leaves and soil.

The rate of solar radiation absorbed by the trees is smaller in the middle of the day. The transpiration of trees increases in the morning with the absorbed radiation after a delay probably caused by the dew deposited on the leaves. Thereafter it follows closely the absorbed radiation.

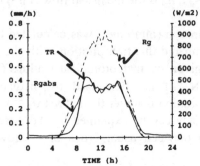

FIG. 2 . Daily course of transpiration (TR, mm h^{-1}) of the trees in relation to global radiation (Rg, Wm^{-2}) and solar radiation absorbed by the trees (Rgabs, Wm^{-2}).

The net radiation absorbed by the trees was also estimated; Fig. 3 shows the comparison between the daily transpiration expressed as a latent heat flux with the first term of the PENMAN equation where the daily net radiation is replaced by an estimation of the absorbed net radiation.

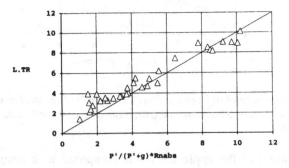

FIG. 3. Relation between daily transpiration expressed as a latent heat flux (L.TR, MJ m^{-2} d^{-1}) and radiative term (MJ m^{-2} d^{-1}) of the PENMAN PET formula where the net radiation is replaced by the net radiation absorbed by the trees (Rnabs); P' is the rate of change of saturated pressure with temperature; g is the psychrometric constant.

The experimental points are pratically all on the 1:1 line. The residual deviations are not significantly dependent on water vapor saturation deficit and wind velocity.

We are elsewhere testing these results are with other configurations of orchards or with drier atmospheric conditions.

Water movements in the trees

The same method was used at Clermont-Ferrand (Central France) in 1989 to compare sap flow between two 2-year-old walnut trees (*Juglans regia* L. cv "Lara"), potted in 200 l containers and subject to normal and drought conditions.

Fig. 4 shows the sap flow in the main trunk and predawn water potential for two walnut trees in normal and drought conditions.

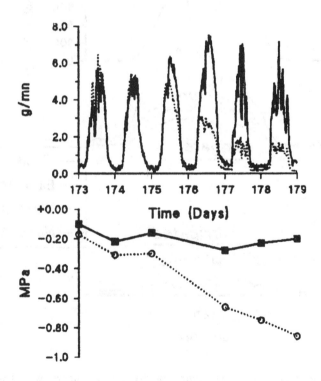

FIG. 4. Sap flow of the main trunk (above) and predawn water potential (below) for two walnut trees: normal condition (continuous line); drought condition (dotted line).

The sap flow in the water-stressed tree becomes lower than in the well watered tree after only two days of water shortage. At that stage the difference in the predawn water potentials is still small. Drought induced the decline of predawn water potential only 4-5 days after irrigation was stopped. The rate of sap flow and therefore transpiration is very rapidly affected by the water stress conditions, while predawn potential reacts more slowly.The heat balance method was also used to compare sap flow within the trunk and the three major branches. Fig. 5 shows variations in global radiation and these sap flows for two trees.

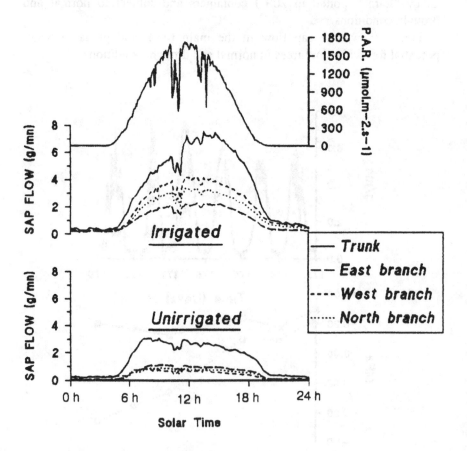

FIG. 5. Daily pattern of sap flow (g/mn) in different parts of two walnut trees under normal and drought conditions in relation to the global radiation (μmol m^{-2} s^{-1}) (day of the year 176).

The pattern of sap flux is similar to that of solar radiation for the irrigated tree. For the drought-treated tree the sap flow increases at the sunset, then it rapidly reaches a limit at around 10 *a.m.* and decreases slowly during the rest of the day. The sap fluxes into the three major branches are synchronous with the trunk sap flow.

For these trees, the spatial variation of the water potential (Fig. 6) shows no difference with respect to orientation and level: solar radiation and water status were similar for the three branches.

In our case, due to the experimental conditions and the form of foliage, the changes of sap flow between the different branches were also synchronous during the day. The relative importance of water flow within each branch is primarily a function of the amount of foliage supported.

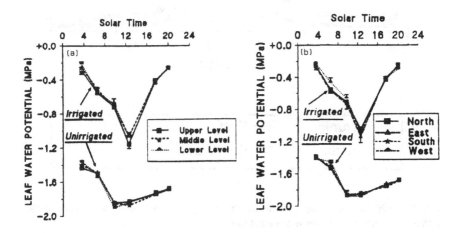

FIG. 6. Spatial variation of the leaf water potential (MPa) for different levels (a) and orientations (b) in two walnut trees under normal and drought conditions (day of the year 180).

The comparison between the trunk and the three major branches was also a test for the accuracy of the method. As can be seen in Fig. 7, the correlation between the flux within the trunk and the sum of fluxes of the three major branches shows some discrepancy due to measurement uncertainty. The sum of fluxes of three major branches was up to 30% greater than the flux within the trunk. These discrepancies would disappear with the last modifications of the geometry of the sap flow sensor and a better insulation of the bases of the trunks and of the containers (Valancogne & Nasr, in this volume; Valancogne, Nasr & Angelocci, data to be published).

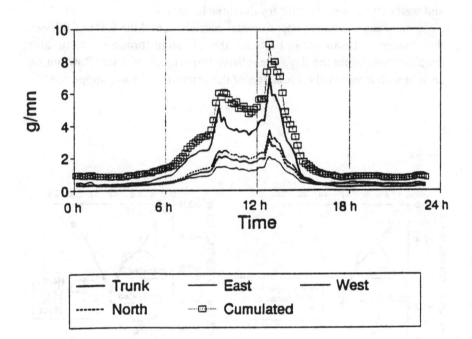

FIG. 7. Daily pattern of sap flow within the trunk as compared with the sum of the fluxes of the major branches in a walnut tree.

The heat balance method for measuring sap flow is sensitive and relatively accurate. Our experiments are showing its usefulness for estimating the water requirements of the trees. Although the heat balance sap flow sensor is difficult to set up by unqualified persons, it is necessary to establish the relations between the water requirements, the climate and the characteristics of the crop.

Added to other measurements of the water status in plants, such as the water potential, this method should allow progress in studies on the water relations in plants, particularly for analysing the different strategies plants develop to adapt themselves to drought conditions.

REFERENCES

Riou, C., Valancogne, C. & Pieri, P. (1989) Un modèle simple d'interception du rayonnement solaire par la vigne. Vérification expérimentale. *Agronomie*, **9**, 441-450.

Sakuratani, T. (1981) A heat balance method for measuring sap flow in the stem of intact plant. *Journal of Agricultural Meteorology*, **37**, 9-17.

Valancogne, C. & Nasr, Z. (1989a) Measuring sap flow in the stem of small trees by a heat balance method. *HortScience*, **24**, 383-385.

Valancogne, C. & Nasr, Z. (1989b) Une méthode de mesure du débit de sève brute dans de petits arbres par bilan de chaleur. *Agronomie*, **9**, 609-617.

Estimating citrus orchard canopy resistance
from measurement of actual and potential transpiration

Y. COHEN, M. FUCHS AND S. MORESHET

Department of Agricultural Meteorology, A.R.O.,
The Volcani Center, Bet Dagan, Israel.

SUMMARY

The relationship between orchard transpiration estimated by a meteorological model or by sap flow in the trunk and canopy conductance was studied in a 17-year-old 'grapefruit' orchard. Low values of canopy conductance during most times of the day are related to the low ratio of potential to actual transpiration in this orchard. High transpiration rate was sustained even when canopy conductance was low, suggesting an important role of atmospheric evaporative demand on transpiration. Under limited soil water availability canopy conductance reached extremely low values, which resulted in reduced transpiration rate. Hourly actual transpiration rates, computed by the model, were well correlated with hourly rates of sap flow in the trunk.

INTRODUCTION

The ratio of actual (Tr) to potential (PTr) transpiration is less than unity because the leaf epidermis offers a resistance to water vapour flow. In a "Marsh" grapefruit (*Citrus paradisi* Macf.) orchard, the ratio was found to be relatively small (0.3) when soil water availability was not limited but it dropped to 0.2 when the soil dried out to about -80 kPa in the main root zone (Cohen, 1991). In the computation of potential transpiration, using a meteorological model, the resistance of the canopy to vapour diffusion is considered zero and this explains the above ratio of measured to potential transpiration. Numerous field and laboratory measurements of stomata have shown the relationship between transpiration and stomatal conductance (Hall & Schulze, 1982). On the other hand, several micrometeorological models have been used successfully to predict evapotranspiration for non-water-stressed vegetation without taking into account the stomatal characteristics (Kanemasu, Stone & Powers, 1976; Ritchie & Jordan, 1972).

The extent to which stomata control transpiration depends largely on the relative magnitudes of stomatal and boundary layer conductance. This approach has been described quantitatively by Jarvis (1985) in terms of a dimensionless decoupling coefficient, which determines the local equilibration of the ambient vapour pressure near the leaf because of transpiration. When the decoupling coefficient is close to unity small changes in stomatal conductance have little effect on transpiration. A large number of studies in citrus have examined the response of stomatal conductance to environmental parameters. Very little was done to relate stomatal conductance to transpiration in the field. Under controlled conditions, stomatal conductance decreased linearly with increasing leaf-to-air water vapour gradient (Hall, Camacho-B & Kaufmann, 1975; Khairi & Hall, 1976). At high soil water potential in a mature citrus orchard, hourly midday conductance was negatively correlated with leaf-to-air vapour pressure difference (Cohen & Cohen, 1983; Maotani & Machida, 1977). Decreasing stomatal conductance with increasing water vapour gradient has been predicted by several empirical transpiration models for trees other than citrus as well as for annual species (Landsberg & Butler, 1980; Avisar et al., 1985; Munro, 1989). In contrast to the sensitivity of stomatal conductance to water vapour gradient, temperature within the range of 20 to 41 °C had only a slight effect on stomata (Hall et al., 1975; Kaufmann & Levy, 1976).

In a controlled environment in which radiant energy was below 200 W m^{-2}, stomatal conductance was linearly related to radiation (Hall et al., 1975) however, under field conditions, canopy conductance was relatively insensitive to changes in increasing radiant energy (Meyer & Green, 1981; Cohen & Cohen, 1983). The present work aims at validation of the above mentioned Tr/PTr ratio in grapefruit which was obtained from sap flow measurements and potential transpiration computed by a meteorological model. Measured canopy conductance will be incorporated into the model for computation of actual transpiration which will then be related to potential transpiration.

MATERIALS AND METHODS

The measurements were carried out in a 17-year-old grapefruit orchard on a sandy soil near Bet Dagan, Israel (32° 01'N, 34° 50'E, 25 m above MSL). Characteristics of the canopy and detailed description of the methodology were described earlier (Cohen, 1991). Leaf area index of the canopy was estimated to be 5.0 to 5.5 based on previous estimations in other citrus orchards in this region (Cohen & Cohen, 1983), and covered approximately 80% of the ground area. Transpiration was determined in eight trees by the heat pulse calibrated method (Cohen, Fuchs & Green, 1981).

Measurements were automated under the control of a 21X data logger (Campbell Scientific, Logan UT, USA). A custom-made multiplexer enabled distribution of the heat pulse and scanning of the heat dissipation. An automatic meteorological station was operated during the measurement days. Global radiation (Kipp and Zonen, Delft, Netherlands), wind speed and direction (Model 014A and 024A met-one, Sunnyvale, CA. USA), temperature and humidity (Model 201 Campbell Scientific) were collected every 7.5 minutes and recorded (CR-7 Campbell Scientific). Canopy height was measured at three points for each location (at the center of the trunk row and 1.5m distance from the trunk row at each side); maximum tree height was 5.0 m. Row width was measured in three points along the sampled location, row direction was north-south. The meteorological input together with orchard characteristics were used to compute the potential transpiration using the Penman model as described by Fuchs, Cohen & Moreshet (1987). Stomatal conductance was determined at intervals of less than one hour, in a set of 48 previously selected leaves.

Two model LI 1600 steady state porometers (Li-Cor, Lincoln NE, USA) were used. Twelve leaves were selected along approximately 5 m of the row in 4 locations as follows: Upper 1.5 m layer in the east side of the row (top east), upper 1.5 m layer in the west side of the row (top west), 1.5 to 3.5 m from the top in the east side of the row (bottom east), 1.5 to 3.5 m from the top in the west side of the row (bottom west). Stomatal conductance, leaf temperature and photosynthetic active radiation (PAR) on the measured leaf surface were automatically recorded from the porometer.

Measurements were collected on two days in which the soil water availability was not limited and during a day in which soil water potential was about -70 KPa in the main root zone.

RESULTS AND DISCUSSION

Diurnal course of leaf conductance showed a slight and gradual decrease with time during the day (Fig. 1). Stomatal measurements were not possible earlier than 0900 h because the leaves were still wet from the dew. In the morning hours, relatively high conductance values were measured on the east side of the row. On the west side, morning values were low (900 to 1000 h), followed by a slight increase. Scattering within a certain set of measurements reflects variations in leaf age, inclination and exposure to the light. Similar responses of stomatal conductance in citrus were reported earlier (Syvertsen, 1982; Camacho-B, 1977; Levy & Kaufmann, 1976) but the values are lower compared with conductance in other species (Denmead & Miller, 1976; Meyer & Green, 1981).

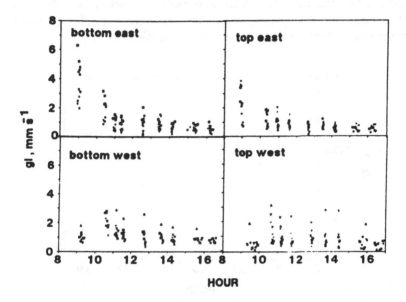

FIG. 1. Diurnal response of stomatal conductance (g_l) of citrus canopy surfaces exposed differently to radiative flux, data are values for single leaves.

The data in Fig. 1 were used for calculating canopy conductance (assuming LAI = 5) and this was incorporated in the transpiration model together with the relevant climatic parameters for computation of actual transpiration. Fig. 2 shows hourly values of actual to potential transpiration during one day of measurements. The ratio dropped from about 0.36 in the morning to 0.28 in the afternoon, which indicates an overestimation of actual transpiration in the morning compared with the afternoon. The overestimation in the morning could be due to the sampled leaves being mostly located in the canopy surface and better exposed to the radiation compared with the entire canopy. For the same reason stomatal closure, due to high water stress, occurred in the sampled leaves earlier than in the entire canopy.

The average daily ratio was 0.31, which is identical to the ratio of sap flow/potential transpiration reported in a previous study for a major part of the irrigation season (Cohen, 1991). In another study, transpiration, (estimated at about weekly intervals by means of soil water balance), varied between 0.19 to 0.60 of pan evaporation but the average value for the entire irrigation season was 0.33 (Kalma & Stanhill, 1969). It may, therefore, be concluded that under non-limited soil water availability the Tr/PTr ratio of 0.3 may be a reasonably accurate factor for estimating citrus orchard transpiration from potential transpiration.

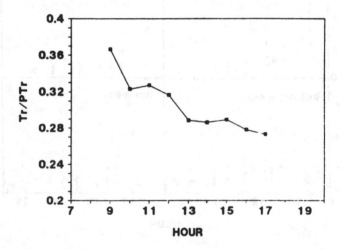

FIG. 2. Transpiration (Tr) to potential transpiration (PTr) ratio in citrus orchard. Each was calculated using the Penman Monteith equation, but for the PTr, canopy conductance was considered zero.

Comparison of the diurnal curve of transpiration computed by the model with that measured by sap flow in the trunk is shown in Fig. 3. The two measurements were reasonably correlated in spite of different sources of error to which each of the methods is exposed. The sap flow curve lagged behind the transpiration curve by one hour or less. The difference may also be attributed to the sampling problems which were discussed above, but may also indicate a phase shift between the two methods. Sap flow is measured in the lower part of the trunk as compared to transpiration from the upper part of the canopy, therefore, the sap flow may lag behind the transpiration from the canopy. Sap flow rate of the two sampled trees was close to the average rate of the other 20 trees measured in the orchard (Cohen, 1991). As may be expected, in view of the limited diurnal response of stomatal conductance shown in Fig. 1, conductance was poorly related to photosynthetic active radiation measured on the leaf surface (Fig. 4).

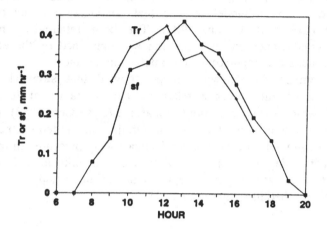

FIG. 3. Diurnal curves of sap flow in the trunk (average of two trees on which stomatal conductance was determined) and transpiration rate, estimated by canopy conductance and relevant climatic parameters.

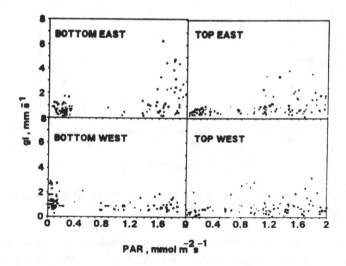

FIG. 4. The response of stomatal conductance (g_l) to Photosynthetic Active Radiation (PAR); data are values for single leaves measured from 0900 h to 1700 h.

The high values at high radiative flux were measured in the morning hours, but during the rest of the day there was in fact no response of stomatal conductance to radiative flux density. The measurements also showed lack of response of stomatal conductance to the gradient of water vapour concentration. These results support earlier findings in Shamouti orange where only 11% of the variations in the measurements were accounted for by a model relating leaf conductance to atmospheric variables (Cohen & Cohen, 1983). The poor relationship between stomatal conductance and transpiration was explained by the effect of internal water status regulating the stomatal conductance.

Measurements in Valencia orange showed that stomatal closure occurred consistently over a relatively narrow range of leaf water potential, which varied according to leaf age (Syvertsen, 1982). Canopy conductance was poorly related to transpiration, computed by the model (Fig. 5). High values of canopy conductance were measured in the early hours of the day when transpiration rate was still low. But when transpiration was at its highest rate, in the noon time, canopy conductance has almost reached its daily minimum value.

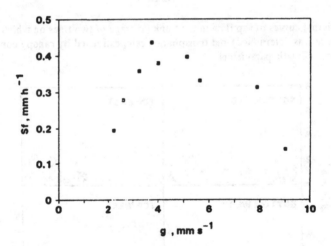

FIG. 5. The relationship between actual transpiration rate (sf), computed by a meteorological model, and canopy conductance (g).

A poor correlation between canopy conductance and evapo-transpiration on a monthly basis was also reported for sweet orange (Van Bavel, Newman & Hilgeman, 1967). In their study, canopy conductance was estimated to be around 2.5 mm s^{-1} throughout the entire year, which is about 50% of the values given in this work for a typical summer day. This poor correlation suggests that low level of stomatal conductance is sufficient to maintain transpiration in citrus even at the highest rate, due to the effect of atmospheric evaporative demands as a driving force for evaporation. There is evidence, however, that if soil water potential drops, stomatal conductance decreases accordingly, and this limits transpiration. Stomatal conductance and transpiration were followed in this study when average soil water potential in the main root zone (0-60 cm) was about -70 KPa.

During the morning hours, stomatal conductance was 40% and transpiration rate 25%, lower than values obtained at the same time under unlimited soil water availability. A similar response of stomatal conductance to soil water potential was obtained in Valencia orange in South Africa (Table 1, taken from Moreshet, S., unpublished data).

Table 1. Frequency analysis of stomatal conductance in Valencia orange under two soil water regimes (data collected on sunlit leaves from 10 h to 17.30 h on a summer day).

Stomatal conductance classes, mm s^{-1}	Frequency analysis for soil water potential of	
	5.8 kPa	> 100 kPa
0.01 - 0.40	3	16
0.41 - 0.80	11	13
0.81 - 1.20	13	8
1.21 - 1.60	7	4
1.61 - 2.00	5	0
2.01 - 2.40	3	0

Stomatal conductance in Valencia was in the same range as in the present study, and significantly dropped in response to soil water stress.

The measurements in this study confirm earlier findings that stomatal conductance in citrus is low compared with other species. It has also been shown that transpiration may be maintained at a high level, regardless of stomatal conductance values in the range of 0.5-2.0 mm s^{-1}. It appears, however, that under soil water stress conditions, stomatal conductance dominates the transpiration rate by dropping to much below such values or even to zero.

ACKNOWLEDGEMENTS

Thanks are due to Mr. Tal Amit, Chief of Zrifin Experiment Farm, The Ministry of Agriculture, who offered the experimental plot and facilities in the farm for the experiment. This research was supported by BARD Grant No. I-944-85, The United States-Israel Binational Agricultual Research Development Fund.

REFERENCES

Avisar, R., Avisar, P., Mahrer, Y. & Bravdo, B.A. (1985) A model to simulate response of plant stomata to environmental conditions. *Agricultural and Forest Meteorology*, **34**, 21-29.

Camacho-B, S.E. (1977) Some aspects of the stomatal behaviour of citrus. *Proceedings International Society Citric*, **1**, 66-69.

Cohen, S. & Cohen, Y. (1983) Field studies of leaf conductance response to environmental variables in citrus. *Journal of Applied Ecology*, **20**, 561-570.

Cohen, Y., Fuchs, M. & Green, G.C. (1981) Improvement of the heat pulse method for determining sap flow in trees. *Plant, Cell and Environment*, **4**, 391-397.

Cohen, Y. (1991) Determination of orchard water requirement by a combined trunk sap flow and meteorological approach. *Irrigation Science*, **12**, 93-98.

Denmead, O.T. & Miller, B.D. (1976) Field studies of the conductance of wheat leaves and transpiration. *Agronomy Journal*, **68**, 307-311.

Fuchs, M., Cohen, Y. & Moreshet, S. (1987) Determining transpiration from meteorological data and crop characteristics for irrigation management. *Irrigation Science*, **8**, 91-99.

Hall, A.E., Camacho-B, S.E. & Kaufmann, M.R. (1975) Regulation of water loss by citrus leaves. *Physiologia Plantarum*, **33**, 62-65.

Hall, A.E. & Schulze, E.-D. (1982) Stomatal responses, water loss and CO_2 assimilation rates of plants in contrasting environments. In *Encyclopedia of Plant Physiology, Vol. 12B, Physiological Plant Ecology II* (ed. O.L. Lange, P.S. Nobel, C.B. Osmon and H. Ziegler), pp. 181-230. Springer-Verlag, New York.

Jarvis, P.G. (1985) Coupling of transpiration to the atmosphere in horticultural crops: The omega factor. *Acta Horticulturae*, **171**, 187-205.

Kalma, J.D. & Stanhill, G. (1969) Transpiration, evaporation and deep drainage losses from an orange plantation. *Israel Journal of Agricultural Research*, **19**, 11-24.

Kanemasu, E.T., Stone, L.R. & Powers, W.L. (1976) Evapo-transpiration model tested for soybean and sorghum. *Agronomy Journal*, **68**, 569-572.

Kaufmann, M.R. & Levy, Y. (1976) Stomatal response of *Citrus jambhiri* to water stress and humidity. *Physiologia Plantarum*, **38**, 105-108.

Khairi, M.M.A. & Hall, A.E. (1976) Temperature and humidity effects on net photosynthesis and transpiration of citrus. *Physiologia Plantarum*, **36**, 29-34.

Landsberg, J.L. & Butler, D.R. (1980) Stomatal response to humidity: implications for transpiration. *Plant, Cell and Environment*, **3**, 29-33.

Levy, Y. & Kaufmann, M.R. (1976) Cycling of leaf conductance in citrus exposed to natural and controlled environments. *Canadian Journal of Botany,* **54,** 2215-2218.

Maotani, T. & Machida, Y. (1977) Studies of leaf diffusion resistance of fruit trees. 1. Methods of measuring leaf diffusion resistance of Satsuma Mandarin trees and factors influencing it. *Journal of Japanese Society of Horticultural Science,* **46,** 1-8.

Meyer, W.S. & Green. C.G. (1981) Comparison of stomatal action of orange, soybean and wheat under field conditions. *Australian Journal of Plant Physiology,* **8,** 65-76.

Munro, D.S. (1989) Stomatal conductance and surface conductance modelling in a mixed wetland forest. *Agricultural and Forest Meteorology,* **48,** 235-249.

Richie, J.T. & Jordan, W.R. (1972) Dryland evaporative flux in a subhumid climate, part 4. Relation to plant water status. *Agronomy Journal,* **64,** 173-176.

Syvertsen, J.P. (1982) Minimum leaf water potential and stomatal closure in citrus leaves of different ages. *Annals of Botany,* **49,** 827-834.

Van Bavel, C.H.M., Newman, J.E. & Hilgeman, R.H. (1967) Climate and estimated water use by an orange orchard. *Agricultural Meteorology,* **4,** 27-37.

Stomatal conductance in tomato responds to air humidity

I. FERREIRA

DER, Instituto Superior de Agronomia,
Universidade Técnica de Lisboa,
Tapada da Ajuda, 1399 Lisboa Codex, Portugal.

SUMMARY

Stomatal conductance and water potential were recorded in an irrigation experiment in which air humidity and soil water were varied. The results suggest that stomata responded to changes in air humidity irrespective of soil water availability when the predawn leaf water potential is above some critical level used for irrigation scheduling.

INTRODUCTION

In previous experimental work the variation of evapotranspiration (ET), stomatal conductance (g_s) and predawn leaf water potential (Ψ) were studied in an irrigated tomato crop in Portugal; the midday stomatal conductance decreased when a Ψ of -0.4 MPa was attained (Katerji, Itier & Ferreira, 1988). Above this limit there was considerable scatter and no clear relationship between these variables existed. In spite of that, ET decreased as soon as one day after irrigation (Itier, Ferreira & Katerji, 1988) and this was associated with a decrease in g_s (Ferreira, 1987). What is the reason for stomatal closure so soon after irrigation? Two hypotheses were considered: 1) local advection reduces the air humidity conditions close to the leaves inducing stomatal closure and 2) the roots send a signal to the shoot (as described by Davies & Mansfield, 1988) and induce stomatal closure.

MATERIALS AND METHODS

Measurements were made in the field in Coruche, Portugal (lat 38°57', long 8°32', alt 30m). Tomato plants (*Lycopersicon esculentum* Mill.) were planted in a loamy soil in rows for furrow irrigation. Four plots (A, B, C and D) were used, each with a 50 m fetch in the center (Fig. 1). After 13

July, plots A and C were left to dry for 17 days but the soil of plot C was covered with a plastic film in such a way that water was flowing above the plastic sheeting but not wetting the soil (Fig. 1). In plots B and D plants were well irrigated but plot D was covered with a plastic film to avoid soil evaporation. Plastic sheeting was installed on the 19th when the soil was nearly dry. Therefore, plot A was expected to have low transpiration and low evaporation from the soil, plot B high transpiration and high evaporation, plot C low transpiration and high evaporation and plot D high transpiration and low evaporation. Water potentials of leaves of the upper part of the canopy were measured using a Scholander pressure chamber. Soil water conditions were very heterogenous and so Ψ was used to represent the average conditions of the soil in the rooting zone. A steady-state porometer (Li-1600, Li-Cor, Inc., Lincoln, Nebraska, USA) was used to measure the abaxial stomatal conductance using leaves from the upper part of the canopy. Measurements of g_s were made between 11:30 and 15:30 and means were calculated from a set of at least 16 individual values per plot.

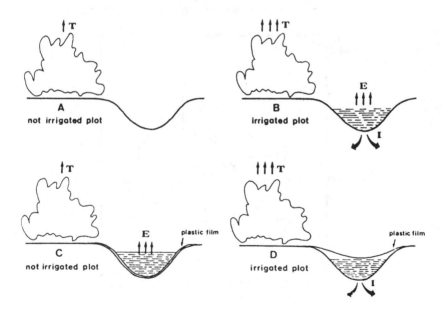

FIG. 1. Schematic representation of the conditions of evaporation and transpiration on different plots (A-D).

Fig. 2 shows the day to day evolution of midday g_s and Ψ in the 4 plots. In plots A and C, Ψ is similar and lower that in the well irrigated plots B and D. In plot A, g_s was significantly less than in the other plots, even plot C which had similar values for Ψ but a high air humidity as consequence of evaporation of water between rows. Fig. 3 shows the relationship between g_s and the leaf-to-air vapour pressure deficit for all the mean hourly values , for all the plots. A single straight line describes this relationship.

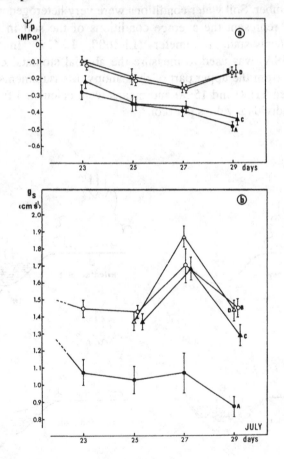

FIG. 2. Predawn leaf water potential (a) and abaxial stomatal conductance (b) of tomato plants, for plots A (solid circles), B (open circles), C (solid triangles) and D (open triangles).

DISCUSSION

These results indicate that the evaporation from a wet soil between rows may alter in an important way the stomatal behaviour of the crop. Soil evaporation influences air humidity near the leaves and changes in humidity appear to be the main reason for the difference between g_s measured on plants from non-irrigated plots A and C. The lower values of g_s are related to higher values of the leaf-to-air vapour pressure deficit independently of predawn leaf water potential.

It seems that when Ψ is higher than some critical level the relation between Ψ and g_s is obscured by the response to air humidity conditions. For $\Psi > -0.5$ MPa no relationship between g_s and Ψ was found unless the measurements corresponded to the same saturation deficit of the air. Thus, during the usual interval between irrigations it appears that g_s is strongly dependent on air humidity.

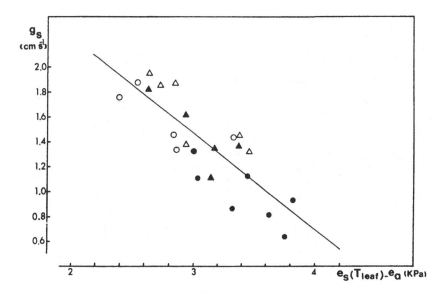

FIG. 3. Abaxial stomatal conductance *vs.* leaf to air vapour pressure deficit for all the mean values corresponding to sunny conditions at noon $\pm$ 1 hour for all the plots (same symbols as in Fig. 2).

ACKNOWLEDGEMENTS

I am grateful to Prof. Joâo Pereira for critical revue of the manuscript and discussion. I thank INIC for its financial contribution to the materials used and INIA for providing the use of facilities at the experimental farm.

REFERENCES

Davies, W.J. & Mansfield, T.A. (1988) Abscisic acid and drought resistance in plants. *ISI Atlas of Science: Animal and Plant Sciences*, 1, 263-269.
Ferreira, M.I. (1987) *Evapotranspiraçäo real. Estudo realizado na cultura do tomate em regiäo de clima mediterrâneo.* Ph.D. Thesis, Instituto Superior de Agronomia, Lisboa.
Itier, B., Ferreira, M.I. & Katerji, N. (1988) Evolution journalière du coefficient de sécheresse entre deux irrigations sur tomate. In *Time Scales and Water Stress* (ed. F. Di Castri, Ch. Floret, S. Rambal & J. Roy) pp. 191-196. Proceed. 5th Int. Conf. Mediterr. Ecosystems, IUBS, Paris.
Katerji, N., Itier, B. & Ferreira, M.I. (1988) Étude de quelques critères indicateurs de l'état hydrique d'une culture de tomate en région semi-aride. *Agronomie*, 8, 425-433.

Water relations of Canarian laurel forest trees

R. LÖSCH

Abt. Geobotanik, Universitätsstrasse 1/26.13,
D-W-4000 Düsseldorf, Germany.

SUMMARY

Water supply capacity of the xylem, water loss avoidance and drought tolerance of leaves of Canarian laurel forest trees were investigated and compared with corresponding data for Mediterranean sclerophylls. Generally, the lauriphyllous species are unable to control their canopy water relations if they are affected by arid conditions. This may be one reason why the Macaronesian laurel forests are restricted to perhumid sites.

INTRODUCTION

The laurel forest of the Macaronesian archipelagos (Canary Islands, Madeira, Azores) is restricted to moist and humid mountain slopes where humidity brought by the trade winds condenses to form clouds (Ceballos & Ortuno, 1976). The tree species of these forests are obviously not competitive at low humidity. This inability to withstand arid conditions may arise from (i) an insufficient xylem capacity for water transport to the crowns, (ii) an ineffective stomatal regulation of transpiration, (iii) a low drought tolerance of the leaves. These functional attributes of the laurel forest tree species were investigated comparatively and compared with pertinent data of Mediterranean species.

MATERIALS AND METHODS

The species are named in the legend of Fig.1. Xylem anatomy and drought tolerance of the leaves were studied with plant material collected in the Anaga Mountains of Tenerife, Canary Islands (28.5° N, 15.9° W). Transpirational water loss was assessed with leaves from greenhouse-grown plants. No differences in structure and function could be recognized between field-collected and cultivated plants. The theoretical

water supply capacity of the twigs was calculated according to the Hagen-
Poiseuille formula (Huber, 1956) using data from microscopic
determinations of vessel numbers and diameters. Weight loss of saturated,
excised leaves under constant ambient conditions (24 °C, 60% r.h.) was
measured continuously for one hour after detachement and was taken as
an integrative measure of the ability to avoid water loss by both stomatal
closure and cuticular resistance. Drought tolerance of leaves was
determined as the critical water saturation deficit beyond which no more
resaturation was possible and which was accompanied by first appearance
of necroses (Bornkamm, 1958). Data for Mediterranean species have
been taken for comparative purposes from the literature (Rouschal, 1939;
Larcher, 1984) or determined on greenhouse-grown plants.

RESULTS

Calculated values of xylem conductivity, vital limits of drought tolerance,
and the measure of drought avoidance are plotted in Fig.1 for each
species into a system of coordinates so that the species less efficient in the
regulation of their water relations are positioned more proximate to the
origin of the axes. The more efficient ones become positioned more
distantly. Drought avoidance of the Mediterranean species (2-6%/h) is
more efficient than that of the laurel forest trees (4-8%/h). Only the
members of the laurel family are comparable in their water supply capacity
with the Mediterranean plants; the other laurel forest species are less
efficient. The drought tolerance of the Mediterranean sclerophylls is
generally higher than that of the lauriphyllous species (50-70% and 30-
45%, respectively).

DISCUSSION

All the data show that the lauriphyllous trees investigated are inferior to
sclerophyllous species in terms of water stress avoidance and tolerance.
Additionally, the water supply to the canopy is limited in some of these
trees by a low-performance xylem system. Even a good capability for
water extraction from the soil (if it exists at all) would be offset by such
an inefficient conducting system. The Macaronesian laurel forest, a relict
of the vegetation which in Tertiary times covered all the land around the
Tethys sea (Cifferi, 1962), turns out to be functionally unable to endure
even moderately arid situations. It has been replaced therefore by more

efficient plants in most of its former range of occurrence and continues to exist only under nearly stress-free conditions.

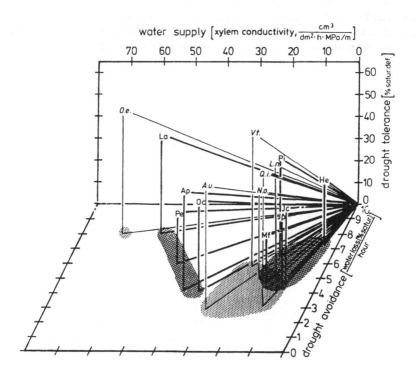

FIG. 1. Calculated xylem water supply capacity, drought tolerance, and water loss avoidance of leaves of Canarian lauriphyllous species and Mediterranean sclerophyllous species. **Mediterranean:** O.e.=*Olea europaea* L., A.u.=*Arbutus unedo* L., V.t.=*Viburnum tinus* L., L.n.=*Laurus nobilis* L., Q.i.=*Quercus ilex* L., N.o.=*Nerium oleander* L. **Canarian:** La=*Laurus azorica* (Seub.) Franco , Ap=*Apollonias barbujana* (Cav.) Bornm., Oc=*Ocotea foetens* (Ait.) Bent & Hook., Pe=*Persea indica* (L.) Spreng., Pi= *Picconia excelsa* (Ait.) DC., Mf=*Myrica faya* Ait., Jc=*Ilex canariensis* Poir., Jp=*Ilex perado* Ait., He=*Heberdenia excelsa* (Ait.) Banks ex Roem. & Schutt.

REFERENCES

Bornkamm, R. (1958) Standortsbedingungen und Wasserhaushalt von Trespen-Halbtrockenrasen (Mesobromion) im oberen Leinegebiet. *Flora* (Jena), **146**, 23-67.

Ceballos, L.F. & Ortuno, F.M. (1976) *Vegetacion y flora forestal de las Canarias occidentales.* Cabildo insular, Sta. Cruz de Tenerife.

Cifferi, R. (1962) La laurisilva canaria: una paleoflora vivente. *Ricerca scientifica* (Roma), **2**, 111-134.

Huber, B. (1956) Die Gefäßleitung. In *Handbuch der Pflanzenphysiologie 3* (ed. W. Ruhland), pp. 541-582. Springer-Verlag, Berlin.

Larcher, W. (1984) *Ökologie der Pflanzen.* Ulmer-Verl., Stuttgart.

Rouschal, E. (1939) Zur Ökologie der Macchien. Der sommerliche Wasserhaushalt der Macchienpflanzen. *Jahrbücher für wissenschaftliche Botanik,* **87**, 436-523.

Watering regime and photosynthetic performance of *Gunnera tinctoria* (Molina) Mirbel.

G. J. CAMPBELL AND B. A. OSBORNE

Botany Department, University College,
Dublin Belfield, Dublin 4, Ireland.

SUMMARY

The influence of water status on photosynthesis and productivity of *Gunnera tinctoria* (Molina) Mirbel. was examined under natural conditions in an experimental garden in Dublin under two contrasting watering regimes. Application of water in amounts that simulated the rainfall conditions of the west of Ireland produced large increases in productivity although there were only small effects on instantaneous rates of net photosynthesis. The lack of a correlation between photosynthetic rate and biomass accumulation is considered. Stomatal conductance was often highest in the morning and subsequently declined throughout the day, irrespective of watering regimes, indicating possible limitations in water transport from the root to the shoot even when plants were supplied with additional water. The significance of these results for a plant restricted to wet, humid environments is discussed.

INTRODUCTION

The recent habitats occupied by the genus *Gunnera* L. are characterised by high rainfall, high humidity conditions in which low temperatures are rare (Osborne *et al.*, 1991; Palkovic, 1974). The global distribution of the genus is also thought to have been more extensive in the past but has decreased dramatically due to a reduction in annual levels of precipitation and an increase in seasonal temperature variation (Osborne *et al.*, 1991; Jarzen, 1980).

Gunnera tinctoria (Molina) Mirbel was introduced into Ireland about 100 years ago as an ornamental shrub and is now naturalised in westernmost regions of the country (Campbell & Osborne, 1990) in areas of high annual rainfall (>1600 mm) and high humidity not unlike its tropical and sub-tropical habitats where rainfall approaches 2000 mm per year (Müller, 1982; Osborne, 1989). There is, however, no evidence of eastward colonisation of habitats in Ireland where rainfall is significantly less than 1600 mm per year. In view of these observations, and because a number of the proposed near relatives of the species are submerged aquatic plants (Arber, 1963) with little tolerance of water deficits, we have examined the effects of water supply on plant performance in an attempt to understand the restriction of the species to high rainfall environments.

MATERIALS AND METHODS

Plants of *G. tinctoria* were collected in November 1985 from a roadside ditch at Curraun Co. Mayo, Ireland (Irish National Grid Reference L810600) and subsequently grown in John Innes potting compost in an unheated greenhouse at University College, Dublin. From this material 18 plants were transplanted to a small experimental garden in May 1986 (Osborne, 1988) . These were subsequently watered regularly for one year and after this period were given two contrasting watering treatments: 9 plants left in normal Dublin rainfall and supplemented with tap water only during very dry periods; 9 plants given 20 litres of water every day when rainfall was less than or equal to 3 mm per day (equivalent to the daily rainfall of *Gunnera* habitats in the west of Ireland). This regime was maintained from March through November which approximates the growing season of the species in Ireland. Monthly diurnal net photosynthesis rates (P_n) were determined from CO_2 exchange in 1989 using a portable infra-red gas analyser (Model LCA-2, Analytical Development Company, Hoddesdon, Hertfordshire, England) and data logger (Type DL2, Analytical Development Company). Measurements were made with a clamp-on leaf chamber (PLC, Analytical Development Company, area=10.00 cm^2) at intervals of approximately 15 minutes throughout the day, on at least three different leaves from three different plants all of the same age (Osborne, 1988).

Maximum potential photosynthesis (P_n max) and photon yields (ϕ) based on incident light (I_i) were estimated for detached leaf discs using measurements of oxygen exchange. Rates of O_2 exchange were measured using a leaf disc electrode (model LD2, Hansatech Ltd., King's Lynn, Norfolk, U. K.) and light source (model LS2H, Hansatech Ltd.) and associated control boxes (models CB1-D and LS2, Hansatech Ltd.) as described by Osborne (1989). Pre-dawn and mid-day leaf water potentials were determined from petioles held in rubber bungs and placed in a pressure chamber (Chas. W. Cook & Son, Birmingham, England). In most cases the intact petioles were too large and petiole cores were used. Where comparisons were possible no difference was found between intact petioles and petiole cores. Water contents and relative water contents were determined from fresh and dry weight measurements using the techniques described by Kramer (1988).

In September 3 plants from each treatment were harvested, separated into their component parts and oven dried at 80 °C for 72 hours. Weights of seed, inflorescence, leaf, petiole, rhizome and roots were determined separately.

Statistical analyses of data were performed using standard t-tests performed on the Apple-Macintosh *Statworks* program.

Climatic data were obtained from the Meteorological Office, Dublin, Ireland.

RESULTS

Fig. 1 shows the water received by both treatments and the temperature experienced in the experimental garden during the growing season. Watered plants received approximately a three-fold increase in H_2O and this was correlated with an approximately six-fold increase in biomass in the watered material (Table 1). Leaf area quotients (also termed leaf area ratio, defined as the ratio of total leaf area to total plant dry weight) show no difference between the two watering treatments. Despite these differences in biomass no significant difference in maximum photosynthetic O_2 exchange rates ($P>0.1$) or photon yields ($P>0.1$) were found (Fig. 2).

TABLE 1. Biomass Accumulation, Leaf Area Quotients and Specific Leaf Area from watered and unwatered treatments obtained from material harvested close to the end of the growing season. Standard errors in brackets.

Biomass/Growth parameter	Watered		Unwatered	
Seeds (kg)	0.50	(0.20)	0.20	(0.02)
Inflorescences (kg)	0.13	(0.60)	0.04	(0.06)
Petioles (kg)	1.10	(0.40)	0.03	(0.02)
Leaves (kg)	2.60	(0.10)	0.02	(0.01)
Rootstock (kg)	8.40	(1.20)	2.0	(0.06)
Roots (kg)	0.90	(0.00)	0.02	(0.06)
Total iomass (kg)	14.30	(2.50)	2.50	(0.50)
Leaf area quotient ($m^2 kg^{-1}$)	0.50	(0.01)	0.05	(0.01)
Specific leaf area ($kg m^{-2}$)	18.82	(3.29)	19.23	(3.85)

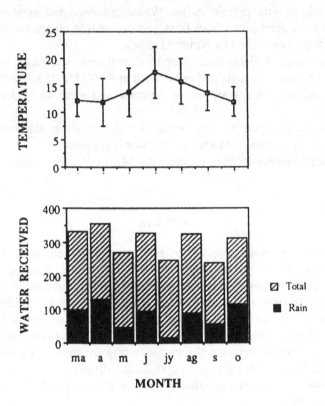

FIG. 1. Seasonal variations in temperature (°C) and water received (1 m⁻²) during the growing season of 1989 in the experimental garden at University College Dublin. 'Rain' depicts actual rainfall while 'Total' is simulated rainfall.

A decline in photosynthetic rate in November (Fig. 3) was evident with both watered and unwatered plants and was probably due to the deterioration of leaf tissue.

CO_2 exchange rates exhibited only a slight seasonal variation between watered and unwatered material with no significance ($P > 0.1$) calculated from the recorded data (Fig. 4). The highest photosynthetic rates were correlated with the highest stomatal conductances during the early morning. Both stomatal conductance and photosynthesis subsequently declined throughout the day. This was particularly evident on warm days during July and August (Fig. 1) when vapour pressure deficits increase during the day and a peak photon flux density occurs around noon (Fig. 4).

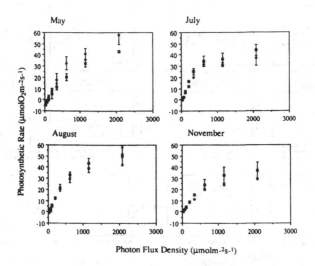

FIG. 2. Light response curves for O_2 exchange measurements of watered (open) and unwatered (closed) material.

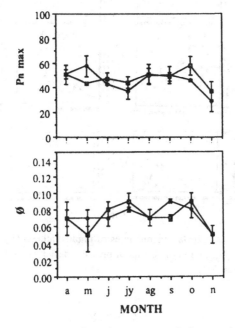

FIG. 3. Monthly measurements of maximum rates of photosynthesis (P_n max, μmol CO_2 m^{-2} s^{-1}) and photon yield based on incident light for watered (open) and unwatered (closed) plants.

252 G.J. CAMPBELL AND B.A. OSBORNE

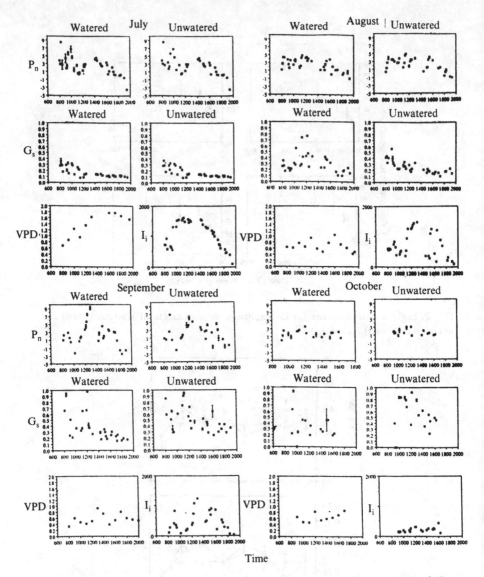

FIG. 4. Diurnal variations in instantaneous rates of photosynthesis (P_n: μmol CO_2 m^{-2}s^{-1}), stomatal conductance (G_s: mol m^{-2}s^{-1}), vapour pressure deficit (VPD: kPa) and incident light (I_i: μmol photon m^{-2}s^{-1}).

DISCUSSION

High stomatal conductances and peak photosynthetic rates in the early morning with a gradual decline during the day is often indicative of extreme water deficits, high temperatures (Roessler & Monson, 1985; Davis, 1987) or declining stomatal conductances in response to increasing vapour pressure deficits (Turner, Schulze & Gollan, 1985). This response is often preceded by a noon depression of CO_2 uptake with a complete or partial recovery during the afternoon (Lange *et al.*, 1975), but this was not observed in this work. On cloudless days during summer 1989, P_n max for *G. tinctoria* occurred during the early-morning hours before 10:00 h and a decline was evident by mid-late morning.

These diurnal trends are similar to those reported by Davis (1987) and Mika & Antoszewski (1972) and are usually associated with plant responses to extreme environmental conditions. Studies on mediterranean sclerophylls suggest that there is an increased tendency for midday stomatal closure as the water potential of the leaves becomes more negative (Tenhunen, Lange & Braun, 1981; Tenhunen, Lange & Jahner, 1982); *G. tinctoria*, however, did not show a marked reduction in either water potential or relative water content (Fig. 5). Similar responses were found by Chaves *et al.* (1987) where the rate of CO_2 assimilation of grapevines was depressed at moderate leaf water deficits or even before an apparent drop in leaf water deficit was seen.

In *G. tinctoria* these responses are evident, irrespective of watering regime, indicating a possible limitation in water transport from the roots to the shoot and/or a localised leaf water deficit. This may be a factor restricting the species to wet, humid environments where constraints on water transport would be less significant. The demonstration of similar photosynthesis and stomatal conductances in both watered and unwatered plants indicates, however, that reductions in photosynthetic capacity, even in high rainfall habitats, are likely.

Only small differences in absolute rates of photosynthesis are found between the two treatments, with watered plants showing a marginally higher rate during the first 2-3 months of the growing season. It is therefore difficult to account for the differences in biomass based on measured rates of photosynthesis. Although differences in biomass can also be accounted for by variations in the allocation of dry mass into photosynthetic structures (Lambers, 1987), this cannot account for the differences in this work as the leaf area quotients (= leaf area ratio) and specific leaf areas are similar (Table 1).

If the differences found between watered and unwatered plants during the early part of the season are maintained throughout the growth period, provisional estimates based on leaf carbon budgets suggest that this could account for the differences in biomass accumulation. This indicates that a more detailed assessment of gas exchange is required, particularly during the early morning and the beginning of the growing season to assess the relationship between gas exchange data and growth accurately. It may, however, be difficult to relate leaf gas exchange data directly with biomass production for a species such as *G. tinctoria*, where a large proportion of the plant mass is found in the rhizome (Osborne *et al.*, 1991).

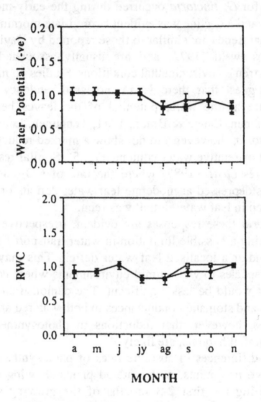

FIG. 5. Water potentials (MPa) and relative water contents (g g^{-1} fresh weight) of watered (open) and unwatered (closed) plant material from the experimental garden during the growing season April-November 1989.

ACKNOWLEDGEMENTS

We wish to thank EOLAS, The Irish Science and Technology Agency, for assistance in funding this project and Eamonn Mahon and Ann Cullen for technical assistance.

REFERENCES

Arber, A. (1963) *Water Plants*. Strauss and Cramer, Leutershausen.

Campbell, G.J. & Osborne, B.A. (1990) *Gunnera tinctoria* (Molina) Mirbel in western Ireland. *Irish Naturalists Journal*, **22**, 223-223.

Chaves, M.M., Harley, P.C., Tenhunen, J.D. & Lange, O.L. (1987) Gas exchange studies in two Portuguese grapevine cultivars. *Physiologia Plantarum*, **70**, 639-647.

Davis, T.D. (1987) Diurnal and seasonal patterns of net photosynthesis by irrigated *Chrysothamnus nauseosus* under field conditions. *Photosynthesis Research*, **11**, 201-209.

Jarzen, D.M. (1980) The occurrence of *Gunnera* pollen in the fossil record. *Biotropica*, **12**, 117-123.

Kramer, P.J. (1988) Measurements of plant water status: Historical perspectives and current concerns. *Irrigation Science*, **9**, 275-287.

Lambers, H. (1987) Does variation in photosynthetic rate explain variation in growth rate and yield? *Netherlands Journal of Agricultural Science*, **35**, 505-519.

Lange, O.L., Schulze, E.D., Kappen, L., Buschbom, U. & Evenari, M. (1975) Photosynthesis of desert plants as influenced by internal and external factors. In *Perspectives of Biophysical Ecology*. (ed. D.M. Gates & R.B. Schmerl), pp. 121-143. Springer-Verlag, Berlin.

Mika, A. & Antoszewski, R. (1972) Effect of leaf position and tree shape on the rate of photosynthesis in the apple tree. *Photosynthetica*, **6**, 381-386.

Müller, M.J. (1982) *Selected Climatic Data for a Global Set of Standard Stations for Vegetation Science*. Dr. W. Junk, The Hague-London.

Osborne, B.A. (1988) Photosynthetic characteristics of *Gunnera tinctoria* (Molina) Mirbel. *Photosynthetica*, **22**, 168-178.

Osborne, B.A. (1989) Effect of temperature on -hotosynthetic CO_2-exchange and slow fluorescence characteristics of *Gunnera tinctoria* (Molina) Mirbel. *Photosynthetica*, **23**, 77-88.

Osborne, B.A., Doris, F., Cullen, A., McDonald, R., Campbell, G. & Steer, M. (1991) *Gunnera tinctoria*: An unusual nitrogen-fixing invader. *Bioscience*, **41**, 224-234.

Palkovic, L.A. (1974) The genecology of *Gunnera* L. in Mexico and Central America. Ph.D. Thesis, Harvard University.

Roessler, P.G. & Monson, R.K. (1985) Midday depression in net photosynthesis and stomatal conductance in *Yucca glauca*. *Oecologia*, **67**, 380-387.

Tenhunen, J.D., Lange, O.L. & Braun M. (1981) Midday stomatal closure in Mediterranean type sclerophylls under simulated habitat conditions in an environmental chamber. 2. Effect of the complex of leaf temperature and air humidity on gas exchange of *Arbutus unedo* and *Quercus ilex*. *Oecologia*, **50**, 5-11.

Tenhunen, J.D., Lange, O.L. & Jahner, D. (1982) The control by atmospheric factors and water stress of midday stomatal closure in *Arbutus unedo* growing in natural macchia. *Oecologia*, **55**, 165-169.

Turner, N.C., Schulze, E.D. & Gollan, T. (1985) The responses of stomata and leaf gas exchange to vapour pressure deficits and soil water content. 2. In the mesophytic herbaceous species *Helianthus annuus*. *Oecologia*, **65**, 348-355.

Water relations and ultrasound emissions in Douglas-fir seedlings infected with xylem pathogens

M. PISANTE,[1] N. MORETTI,[1] AND S. FRISULLO [2]

[1]Dipartimento di Produzione Vegetale, Università della Basilicata,
via N. Sauro 85, 85100 Potenza, Italy.
[2]Dipartimento di Patologia vegetale, Università di Bari,
sede di Foggia, Foggia, Italy.

SUMMARY

Water relations and ultrasound acoustic emissions were measured before, during and after a cycle of water stress, in Douglas-fir (*Pseudotsuga menziesii* (Mirb.) Franco) seedlings infected with two xylem pathogens (*Phomopsis occulta* and *Diplodia pinea*). After 150 days from the inoculation no significant differences in water relations parameters and ultrasound emissions were found between infected and control plants.

INTRODUCTION

Modifications of climate may create more favourable conditions for the development and spread of fungal pathogens. On the other hand, plants may become more susceptible to pathogens if stress conditions (for instance water stress) occur as a consequence of climate changes (Palti, 1981).

Recently, two fungi species (*Phomopsis occulta* Trav. and *Diplodia pinea* (Desm.) Kickx) were found to infect the cambium and the xylem of Douglas-fir (*Pseudotsuga menziesii* (Mirb.) Franco) trees grown in Southern Italy (Basilicata); this observation represents a new record for this region.

Phomopsis occulta causes a definite die-back of young branches and shoots, a typical stem girdling where chromatic alterations are evident with a definite canker of limited growth on the trunk (Wilson & Hahn, 1929). *Diplodia pinea,* chiefly important as a cause of die-back in pines, is recognized as causing a "blight" of Douglas-fir seedlings (Peace, 1962).

As microorganisms which probably infect both the parenchyma and the xylem, these fungi may determine a reduction of hydraulic conductivity of stem and branches, either by physically blocking the xylem conduits or by promoting the formation and spread of gaseous emboli within the conduits (Zimmermann, 1983; Tyree & Sperry, 1989). In this context, research was undertaken to ascertain the possibility of infecting young Douglas-fir seedlings with fungi isolated from older trees, and to evaluate the effects of the infection on xylem water potential, xylem volume fractions, whole plant transpiration and ultrasound acoustic emissions before, during and after a period of water stress.

MATERIALS AND METHODS

Fungi were isolated from 25-year-old trees growing near Potenza (Southern Italy, 40° 38'N, 15° 46'E, 650 m a.s.l.). Isolates were studied as colonies grown on potato-sucrose-agar plates kept at 25 ºC for 30-40 days, and their vitality was assessed.

Fungi were inoculated on February 23, into the stem of 4-year-old potted seedlings, growing in a peat and sand mixture. To inoculate the wood a small "window" of bark was removed and the infected growing medium was put into contact with the xylem. Afterwards the stem was covered with cotton and rubber tape. Five treatments, of 30 plants each, were defined: a) plants inoculated with *Phomopsis occulta*; b) plants inoculated with *Diplodia pinea*; c) plants inoculated with both fungi; d) plants treated with the sterile growing medium; e) control plants. An isolation trial from the xylem of these plants was carried out 60 days after inoculation.

Physiological measurements were performed on three plants per treatment. Different series of plants were used for destructive and non-destructive measurements. Measurements were performed before, during and after a period of water stress, which was imposed by interrupting water supply to plants for one month between April 25 and May 24 1990. Water potential was measured with a pressure chamber (PMS, Corvallis, Oregon, USA) on three twigs per plant. Three plants per plot were sampled. At different dates a different patch of plants was used for destructive measurements. Transpiration was measured by weighing the pots in sealed plastic bags every two hours to the nearest 0.1 g. Xylem volume fractions (water, solid and gas fractions) were measured in accordance with the relationships outlined by Whitehead & Jarvis (1981). For each plant three 30 mm long stem segments were sampled above the "inoculation window" and were stripped of their bark. Volumes were measured as immersed weight in distilled water, and weights were determined to the nearest 0.1 mg. Ultrasound acoustic emissions (UAE)

258   Water relations in seedlings infected with xylem pathogens

were determined using a PAC (Physical Acoustic Corporation, NJ, USA)
115I ultrasound transducer, whose signals were amplified 75 dB using the
PAC 4615 Drought Stress Monitor.

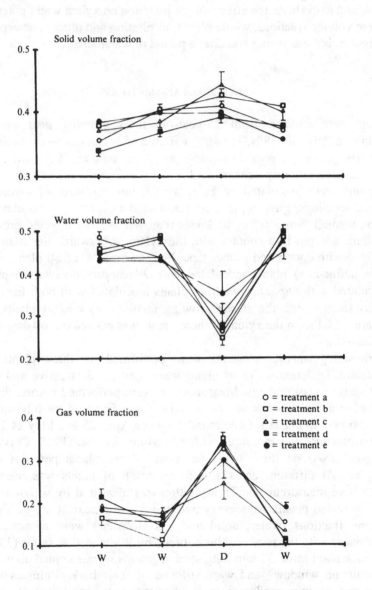

FIG. 1. Variations of the xylem volume fractions during the experiment; W = well
watered plants; D = droughted plants. Bars are standard errors.

The transducer was clamped to the xylem after removing the bark in a small window close to the inoculation point. The xylem was coated with silicon grease to prevent wood dehydration. Counts were made for five minutes on each plant at 2 hour intervals.

RESULTS AND DISCUSSION

Isolation trials carried out after 60 days demonstrated that artificial infections gave positive results on all the plants inoculated with *Phomopsis occulta* or *Diplodia pinea*. In the case of plants inoculated with both fungi positive results were obtained in 60 out of 100 cases. The xylem water volume fraction was 0.40-0.50 at the beginning of the experiment (well-watered plants) and between 0.25 and 0.40 after a month without irrigation. After rewatering, the plants were able to recover their initial xylem water content. No significant differences were found between the treatments (Fig. 1).

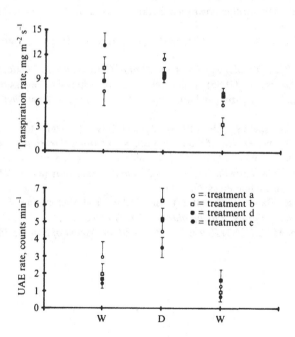

Fig. 2. Variation of transpiration rate and ultrasound acoustic emission during the experiment; W = well watered plants; D = droughted plants. Bars are standard errors.

Well-watered plants displayed a water potential of -1.0 MPa; the water potential fell down to -2.5 / -3.0 MPa after the cycle of water stress and recovered up to -1.3 MPa 3 days after rewatering. Treatments did not show different patterns of drying or recovering. Variations of transpiration rate and UAE in the course of the experiment are shown in Fig. 2.

In accordance with other findings (Peña & Grace, 1986; Jones & Peña, 1986; Borghetti, Raschi & Grace, 1989), UAE increased significantly in droughted plants, which exhibited low transpiration rates (Fig. 2), but no significant differences were found between treatments. Even though no physiological response was detected 150 days after inoculation, long term effects should still be evaluated.

REFERENCES

Borghetti, M., Raschi, A. & Grace, J. (1989) Ultrasound emission after cycles of water stress in *Picea abies*. *Tree Physiology*, **5**, 229-237.

Jones, H.G. & Peña, J. (1986) Relationship between water stress and ultrasound emission in apple (*Malus domestica* Borkh.). *Journal of Experimental Botany*, **37**, 1245-1254.

Palti, J. (1981) *Cultural Practices and Infectious Crop Disease*. Springer-Verlag, Berlin.

Peace, T.R. (1962) *Pathology of Trees and Shrubs*. Clarendon Press, Oxford.

Peña, J. & Grace, J. (1986) Water relations and ultrasound emission of *Pinus sylvestris* L. before, during and after a period of water stress. *New Phytologist*, **103**, 515-524.

Tyree, M.T. & Sperry, J.S. (1989) Vulnerability of xylem to cavitation and embolism. *Annual Review of Plant Physiology and Molecular Biology*, **40**, 19-38.

Whitehead, D. & Jarvis, P.G. (1981) Coniferous forest and plantations. In *Water Deficits and Plant Growth*. Vol. VI. *Woody Plants Community* (ed. T.T. Kozlowski), pp. 49-152. Academic Press, New York.

Wilson, M. & Hahn, G.G. (1929) The history and distribution of *Phomopsis pseudotsugae* in Europe. *Phytopathology*, **19**, 979-992.

Zimmermann, M.H. (1983) *Xylem Structure and the Ascent of Sap*. Springer-Verlag, Berlin.

Diurnal fruit shrinkage: a model

A. BERGER[1] AND G. SELLES [2]

[1] *CNRS - CEFE BP 5051. 34033 Montpellier Cedex, France,*
[2] *Comision Nacional de Riego. Teatinos 50 / 5to Piso, Santiago, Chile.*

SUMMARY

Changes in the diurnal pattern of trunk and fruit diameter in field conditions suggests a model of fruit shrinkage more complex than the well known radial transfer model used to explain trunk shrinkage. In this model the fruit transpiration and the phloem flux play an important role.

INTRODUCTION

Some fruits show reversible shrinkage on a daily basis directly linked to their water balance. The recognized interpretation of this process based on a reverse flux of water from the fruit to the plant in response to the transpirational pull of the foliage (Kozlowski, 1972, 1982) is similar to that used to explain diurnal trunk shrinkage (Landsberg, Blanchard & Warrit, 1976; Powell & Thorpe, 1977). This interpretation based on a network of hydraulic resistances and capacitances, and a single water flux between "bark" and xylem (Fig. 5A) gives a good explanation of the time lag between xylem water potential and trunk shrinkage. Our results show that for fruits, the model of reverse flux between fruit and xylem stem is not valid particulary because it cannot explain the delay between fruit shrinkage and changes in xylem water potential.

MATERIALS AND METHODS

Experimental site and plant material

The experiment was carried out in a five-year-old peach orchard (*Prunus persica* L. cv. Fire-Red on GF 305 rootstock). This old variety was chosen because the growing curve of the fruit presents 3 stages (stage I, until end of May: high growth rate; stage II, from beginning of June to mid-July: slow growth rate; stage III, from mid-July to mid-August: high growth rate).

Treatments

The trees were sprinkler irrrigated and two irrigation treatments were applied: "wet treatment", in which a section of the orchard was irrigated at 100% of its maximal evapotranspiration (MET) and "dry treatment" in which a section of the orchard was irrigated with half the amount of water. MET was calculated from the Brochet-Gerbier potential evapotranspiration equation (Brochet & Gerbier, 1972) corrected by a crop coefficient recommended for peach tree irrigation in the South of France. For details see Garnier & Berger (1986).

Measurements

Tree water potential was measured, using a pressure chamber (PMS Instrument Co., Corvallis, Oregon, USA), on leaves enclosed two hours prior to measurement in a cellophane bag covered with aluminium foil. Preliminary measurements have shown that two hours was a sufficient time for recovery of xylem water potential. This measurement gives an estimate of the water potential in the xylem of the shoot, at the point of attachment of the petiole to the leaf. (Begg & Turner, 1970). Water potential related to fruit shrinkage was measured on leaves close to the fruit peduncle (Ψ_B, branch water potential), water potential related to trunk shrinkage was measured by the same way on previously selected young stems growing directly on the trunk (suckers) (Ψ_T trunk water potential). We decided to separate trunk and shoot water potential because a previous study showed a difference between them as great as 0.4 to 0.5 MPa (Garnier & Berger, 1987). One tree in each treatment was equipped with 2 linear variable displacement transducers (LVDT), each one mounted on an invar alloy frame to reduce the effect of temperature changes (Garnier & Berger, 1986). One of the transducers was placed on the main trunk, 10-30 cm above ground; the other one on a fruit growing on a south facing main branch. The epidermal conductance of fruits was measured with a steady state porometer (Li-1600, Li-Cor, Lincoln, Nebraska, USA). From the conductance values obtained by this way, the fruit transpiration was calculated using the method described by Jones & Higgs (1982). The instantaneous values obtained in this way were integrated over the total shrinkage period to obtain the values of total transpiration (Table 1).

RESULTS

Diurnal trend of water potential

Fig. 1 shows the diurnal trend of water potential in the trunk (Ψ_T) and water potential in a branch near the fruit (Ψ_B) related to climatic conditions. In both cases water potential falls to a more or less constant value between 0900 h and 1300 h, then increases slowly.

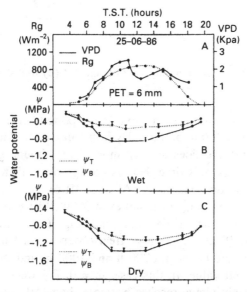

FIG. 1. Diurnal course of climatic conditions (A), trunk (Ψ_T) and branch (Ψ_B) water potential for wet (B) and dry (C) treatment. VPD = vapour pressure deficit, Rg = global radiation, PET = potential evapotranspiration.

Diurnal trend of diameter changes

Fig. 2 presents the diurnal trend of trunk and fruit diameter at stage II of the fruit growth curve. Three parameters characterise the curves: 1) Sh = maximum daily shrinkage; 2) G = daily growth; 3) Δh = duration of the shrinking period. Fruit growth rate is very high (330 μm for the wet treatment and 120 μm for the dry treatment).

The duration of the trunk shrinking period is always greater than 10 hours while the same parameter for fruit shrinkage is always less than 9 hours.

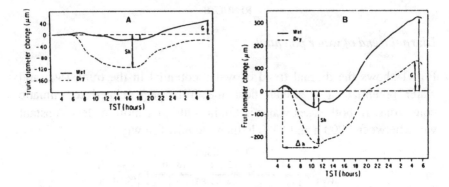

FIG. 2. Daily course of (A) trunk diameter change and (B) fruit diameter change on June 25.

At stage I of the fruit growth curve (not reported here) the growth rate is very high, about 400 to 500 μm day^{-1}, the maximum daily shrinkage is very low (30 μm) and does not last more than 4 hours, generally being one or two hours. At this time there is no significant difference in water stress between the two treatments. At stage III of the fruit growth curve (G = 1 mm) fruit shrinkage does not exist for well irrigated trees (Fig. 3) while the trunk presents a maximum shrinkage (not presented here) of between 20 and 40 μm. The only effect observed on fruit is a slowing down of growth rate between 0700 h and 1000 h, and then growth rate recovers the initial value. In this case Δh = 0. For trees on the dry plot maximum daily shrinkage reaches no more than 150 μm with a growth rate G = 600 μm. The duration of the shrinking period is about four hours.

Whatever the growth stage of the fruit, the trunk invariably recovers some hours later than the trunk water potential, however, the fruit recovers some hours before branch water potential. The very good linear correlation between trunk water potential and branch water potential indicates clearly that there is no important time lag between these two potentials. So, the great time lag between trunk and fruit shrinkage cannot be explained by a time lag in water potential between trunk and the branches.

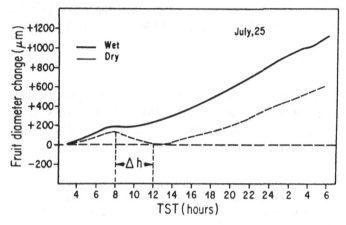

FIG. 3. Daily course of fruit diameter changes on July 25 (stage III).

Effect of growth rate on duration of shrinkage

From the results above it seems that the duration of the fruit shrinkage is affected by the growth stage of the fruit. Plotting Δh against G, for all growth stages, shows clearly that the greater the growth rate, the shorter the shrinking time (Fig. 4A, wet treatment, and Fig. 4B, dry treatment).

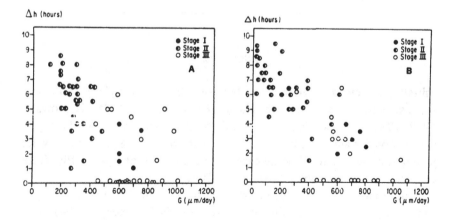

FIG. 4. Relationship between duration of the shrinking phase (Δh) and daily growth rate (G) of the fruit. A: wet B: dry treatment.

There is little difference in fruit growth rate between wet and dry treatment except at stage III, indicating that the dry treatment has little effect on fruit growth, but on the other hand the trunk growth rate is strongly affected after the end of stage I (Selles & Berger, 1990).

Fruit transpiration

The results in Table 1 show that whatever were the growth stage or the irrigation treatment the total transpiration (T) is always greater than volume variation (V). This precludes existence of a global reverse flux between fruit and tree.

TABLE 1. Cumulated transpiration (T, cm³) and fruit volume change (V, cm³) between beginning of shrinkage and time of maximum shrinkage.

Treatment	Stage	T	V
WET	II	2.10	0.81
DRY	II	1.60	1.10
WET	III	2.36	0.46
DRY	II	1.21	1.00
WET	III	2.80	0.00
DRY	III	1.90	0.49

DISCUSSION

Initially, a reverse xylem flux between fruit and xylem stem was suggested to explain fruit shrinkage (Bartholomew, 1926; Kozlowski, 1972, 1982), but measurements of this flux are few and contradictory to the observed water potential gradients (Yamamoto, 1983). In fact, water potential gradients are in some cases in the direction of a water flux from the stem to the atmosphere through the fruit.

Fruit transpiration has been assumed to have an important role as a cause of fruit shrinkage (Jones & Higgs, 1982; Tromp, 1984). Our transpiration measurements strongly support this assumption. The prime importance of fruit transpiration to explain the shrinkage of peach also seems supported by the fact that at *contrario,* for fruits such as tomato with low epidermal water vapor conductance (Nobel, 1975), the diurnal phase never presents a shrinkage but only a growth rate reduction (Ehret & Ho, 1986). So, on peach the assumption of a reverse flux from fruit to stem to explain the shrinkage does not seem reasonable. The initial observation that a branch separated from the plant wilts more slowly if it bears fruits, indicating a water transfer from fruit to leaves, is not a proof of a such transfer in the intact plant. Cutting the branch suppresses water transfer from the soil and reverses the gradient between fruit and leaves.

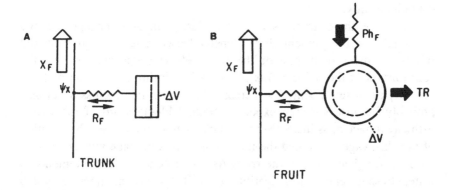

FIG. 5. Model of trunk (A) and of fruit (B) shrinkage. Ψ_x = trunk or branch water potential, X_F = trunk or branch xylem flux, R_F = radial flux, ΔV = volume change, Ph_F = phloem flux, TR = fruit transpiration.

The time lag between trunk shrinkage and water potential drop in the xylem of the trunk is generally explained by a radial water transfer between bark and xylem (Fig. 5A). The water balance of the bark is given by the relation $\Delta V = R_F \, \Delta t$ where ΔV = volume change, R_F = radial flux and Δt = increment of time. The early recovery, in the field, of fruit diameter before xylem water potential recovery does not support such a model. We suggest that the fruit recovers water not only by the radial pathway (xylem of the peduncle) but also and mainly by the phloemic pathway, the relative part of the water flux by this pathway becoming more and more important as the growth rate of the fruit increases. The phloemic water flux from leaves to fruit increases as the photosynthates are accumulating in leaves in the morning and reaches, before noon, such a level as to just compensate the water loss by fruit transpiration. The importance of phloemic pathways has recently been suggested by calculations (Ho & Grange, 1987) indicating that on tomatoes only 15% of the water flux to the fruit could be by xylem, the complement corresponding to the water volume needed for phloemic transport of photosynthates. So, the explanation of fruit diurnal shrinkage needs a model more complex than the model generally used to explain trunk shrinkage. For fruits all the fluxes potentially acting have to be taken into account to explain the daily trend of the water balance (direct and reverse radial flux, phloem flux, transpiration) (Fig. 5B). The water balance of the fruit is given by the relation $\Delta V = (R_F + Ph_F + TR)\Delta t$, where Ph_F is the phloem flux.

Therefore, the characteristics of the diurnal diameter changes depend essentially on the phloem flux intensity between leaves and fruit and finally on the growth rate of the fruit. We can also suggest that the contribution of the phloem flux is increasing during maturation of fruit, not only because phloem flux increases as growth rate increases but also probably because we may expect a gradual blocking up of the xylem pathway in peach, as in some other fruits such as grapes. So, the duration of the shrinkage phase and the maximun daily shrinkage will depend on the physiological stage of the fruit. As a consequence, measurements of water stress in relation to irrigation of fruit trees, as sometimes suggested (Huguet, 1985), should be done carefully (Selles & Berger, 1990) due to a possible seasonal progression of physiological characteristics of the fruit.

REFERENCES

Bartholomew, E.T. (1926) Internal decline of lemons. III. Water deficit in lemons fruit caused by excessive leaf evaporation. *American Journal of Botany,* **13**, 102-117.

Begg, J.E. & Turner, N.C. (1970) Water potential gradients in field tobacco. *Plant Physiology,* **46**, 343-346.

Brochet, P. & Gerbier, N. (1972) Une méthode pratique de calcul de l'évapotranspiration potentielle. *Annales Agronomiques,* **23**, 31-49.

Ehret, D.L. & Ho, L. (1986) Effects of osmotic potential in nutrient solution on diurnal growth of tomato fruit. *Journal of Experimental Botany,* **37**, 1294-1302.

Garnier, E. & Berger, A. (1986) Effect of water stress on stem diameter changes of peach trees growing in the field. *Journal of Applied Ecology,* **23**, 193-209.

Garnier, E. & Berger, A. (1987) The influence of drought on stomatal conductance and water potential of peach trees growing in the field. *Scientia Horticulturae,* **32,** 249-263.

Ho, L.C. & Grange, R.I.(1987) Diurnal tomato fruit growth in relation to phloem transport. *Abstracts of XIV International Botanical Congress* Berlin (West), **1-19,** pp. 28.

Huguet, G. (1985) Appréciation de l'état hydrique d'une plante à partir des variations micrométriques de la dimension des fruits et des tiges au cours de la journée. *Agronomie,* **5,** 733-741.

Jones, H.G. & Higgs, K.H. (1982) Surface conductance and water balance of developing apple (*Malus pumila* Mill.) fruits. *Journal of Experimental Botany,* **132,** 67-77.

Kozlowski, T.T. (1972) Shrinking and swelling of plant tissues. In *Water Deficit and Plant Growth,* Vol III (ed. T.T.Kozlowski), pp. 1-64. Academic Press, New-York.

Kozlowski, T.T. (1982) Water supply and tree growth. Part I: Water deficits. *Forestry Abstracts,* **43,** 57-95.

Landsberg, J.J., Blanchard, T.W. & Warrit, B. (1976) Studies on movement of water through apple trees. *Journal of Experimental Botany,* **27,** 579-596.

Nobel, P.S. (1975) Effective thickness and resistance of the air boundary layers adjacent to spherical plants parts. *Journal of Experimental Botany,* **26,** 120-130.

Powell, D.B. & Thorpe, M.R. (1977) Dynamic aspects of plant-water relations. In *Environmental Effects on Crop Physiology* (ed. J.J. Landsberg & C. Cutting), pp. 259-277. Academic Press, New York.

Selles, G. & Berger, A. (1990) Physiological indicators of plant water status as criteria for irrigation scheduling. *Acta Horticulturae,* **278,** 87-99.

Tromp, J.P. (1984) Diurnal fruit shrinkage in apple as affected by leaf water potential and vapor pressure deficit of the air. *Scientia Horticulturae,* **22,** 81-87.

Yamamoto, T. (1983) Models of water competition between fruits and leaves on spurs of "Bartlet" pear trees and its measurement by heat-pulse method. *Scientia Horticulturae,* **20,** 241-250.

Analysis of pressure-volume curves by non-linear regression

M.N. ANDERSEN,[1] C. R. JENSEN[2] AND R. LÖSCH[3]

[1] *Dep. Soil Tillage, Soil Physics and Irrigation, Danish Research Service for Plant and Soil Science, Flensborgvej 22, DK-6360 Tinglev, Denmark.*
[2] *Dep. Agr. Sci., Section of Soil and Water and Plant Nutrition, The Royal Veterinary and Agricultural University, Thorvaldsensvej 40, DK-1871 Frederiksberg C, Denmark.*
[3]*Abt. Geobotanik, Uni Düsseldorf, Gebäude 26.13/U1, Universitätsstr. 1, D-4000 Düsseldorf, Germany.*

INTRODUCTION

Pressure-volume (PV) isotherms are valuable tools for analyzing plant water status and water transport in the soil-plant-atmosphere continuum. Here we present a new method for the statistical analysis of PV-data by a non-linear regression procedure and derive a mathematical expression for the apoplastic water content.

MATERIALS AND METHODS

The material was drawn from a larger field study on the leaf water relations characteristics of barley (*Hordeum distichum* L. cv. Gunnar) supplied with different amounts of potassium and water. We measured PV-isotherms for leaf No 6 (second prior to the flag leaf) with a pressure chamber. The PV-isotherm (Fig. 1A) may be divided into a nearly linear part below the turgor loss point and an exponential part above the turgor loss point. The nearly linear relation between the osmotic potential (Ψ_π) and the inverse relative water content (R^{-1}) is given by Equation 1, based on the solute content of the symplast (Tyree & Richter, 1982):

$$\Psi_\pi = 1 / (a + bR) \qquad \text{Equation 1}$$

where $a = V_a / (\Phi R^* TN)$ and $b = -(V_o + V_a) / \Phi R^* TN$.

In Equation 1, Φ is an osmotic coefficient, which together with V_a in this treatment is assumed to be constant during dehydration.

Above the turgor loss point we extrapolated Ψ_π in the positive turgor region by Equation 1 and calculated the turgor potential Ψ_p as: $\Psi_p = \Psi_l - \Psi_\pi$ from the observed values of the leaf water potential, Ψ_l. Assuming an exponential relation exists between Ψ_p and the inverse symplastic water fraction R_s^{-1} (Stadelmann, 1984) the turgor can be described as:

$$\Psi_p = \Psi_{p\,(max)}\, e^{\,\beta\,[R_s^{-1} - (1-R_a)^{-1}]} \qquad \text{Equation 2}$$

where β, the sensitivity factor of elasticity, relates exponential changes in turgor to changes in R_s. The parameters a and b of Equation 1 and β of Equation 2 were estimated by the non-linear regression procedure PROC NLIN (SAS Institute, 1988).

To determine R_a we measured the water potential of leaf No 7 (prior to the flag leaf) by dewpoint hygrometry on living leaf disks previously frozen in liquid nitrogen (Fig. 1B). Following the expression given by Tyree & Richter (1982) R_a may be derived from the osmotic potentials of killed (Ψ_π^k) and living (Ψ_π) leaf disks at a given relative water content (R) assuming that the osmotic potential of the apoplastic water is negligible:

$$R_a = (1 - \Psi_\pi^k / \Psi_\pi)\, R \qquad \text{Equation 3}$$

List of symbols

Ψ_l : leaf water potential; Ψ_π: osmotic potential; R: relative water content; V : amount of symplastic water; V_a : amount of apoplastic water; V_0 : amount of symplastic water at $R = 1$; R^* : the gas constant; T: absolute temperature; N : sum of number of moles of all solute species in the symplast; Φ : osmotic coefficient; β : sensitivity factor of elasticity; R_s : symplastic water fraction; R_a : apoplastic water fraction; Ψ_p : turgor potential; $\Psi_{p(max)}$: maximum turgor potential ($-\Psi_\pi$, at $R = 1$); Ψ_π^k : osmotic potential of killed leaf material.

RESULTS AND DISCUSSION

Accurate estimates were obtained for the relations between relative water content and leaf water potential, leaf osmotic potential, and leaf turgor, as further outlined by Andersen, Jensen & Lösch (1991). We used the non-linear PROC NLIN procedure, taking the presence of apoplastic water into account; this causes non-linearity when Ψ_π is estimated by regression of Ψ_π versus $1/R$ (Equation 1). Similarly, the apoplastic water was taken into account when Ψ_p was estimated by PROC NLIN from the relationship between Ψ_p and $1/R$ (Equation 2). A mathematical expression for the apoplastic water fraction (R_a) was derived and measurements (Fig. 1B) gave a value of 0.15 for R_a at $R=1$.

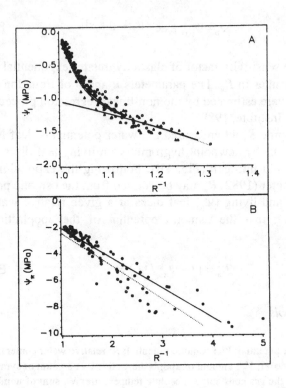

FIG. 1. (A) The relation between Ψ_l and R^{-1} obtained from PROC NLIN estimates. Filled circles and dashed curve represent a treatment given 50 kg K/ha; triangles and fully drawn curve, 200 kg K/ha. The data are from six leaves for each treatment. (B) The relation between Ψ_π and R^{-1} for living and killed leaf tissue. Open circles and dashed line, living material. Filled circles and fully drawn line, killed material. The data are from six leaves for living and killed tissue, respectively.

It has been customary to transform data for Ψ_π and Ψ_p to fit linear models. However, this may distort the error structure and lead to erroneous estimates of Ψ_π and Ψ_p when linear regression analysis is applied.
Finally, the analysis we describe permits data and estimated regression curves to be presented in a readily understood diagram (Fig. 1). Richter

(1978) presents untransformed data and discusses further advantages.

REFERENCES

Andersen, M.N., Jensen, C.R. & Lösch, R. (1991) Derivation of pressure volume curves by a non linear regression procedure and determination of apoplastic water. *Journal of Experimental Botany*, **42**, 159-165.

Richter, H. (1978) A diagram for the description of water relations in plant cells and organs. *Journal of Experimental Botany*, **29**, 1197-1203.

SAS Institute Inc. (1988) *SAS/STAT Users Guide*, Release 6.03 Edition, Cary, NC.

Stadelmann, E.J. (1984) The derivation of the cell wall elasticity function from the cell turgor potential. *Journal of Experimental Botany*, **35**, 859-868.

Tyree, M.T. & Richter, H. (1982) Alternate methods of analyzing water potential isotherms: Some cautions and clarifications. II. Curvilinearity in water potential isotherms. *Candian Journal of Botany*, **60**, 911-916.

Determination of the amount of apoplastic water and other water relations parameters in conifer needles

K. GROSS[1] AND W. KOCH [2]

[1]*Institute of Silviculture, Univ. of Freiburg, Bertoldstr. 17, D-7800 Freiburg,* [2]*Institute of Forest Botany, Univ. of München, Amalienstr. 52, D-8000 München 40, Germany.*

SUMMARY

A method for determining the amount of needle apoplastic water and bulk osmotic pressure in the symplast of spruce needles (*Picea abies* [L.] Karst.) is presented. The method is based upon a combined use of the pressure volume analysis of whole shoots and microcryoscopy of sap pressed from needles after being frozen in liquid nitrogen.

INTRODUCTION

Pressure-volume (PV) analysis (Tyree & Hammel, 1972) enables the determination of several plant water relations parameters. The measured values are "bulk tissue averages", which characterize the plant tissues better when they are uniform (*i.e.* only leaves). In plant species with small leaves, for example conifers, PV analysis can only be performed on shoots which consist of leaves, wood and bark. The resultant water relation parameters then represent a complex tissue; this can be disadvantageous in studying the physiology of leaves or other plant parts.

Through the combined use of PV analysis of spruce shoots and capillary microcryoscopy of sap pressed from needles, determinations of the most important water relations parameters, needle apoplastic water content and bulk osmotic pressure in the needle symplast, were possible.

MATERIALS AND METHODS

Pressure-volume analysis

In late summer 1987, water potential isotherms from different sized shoots of a 25-year-old, 16 meter tall Norway Spruce (*Picea abies* [L.] Karst.) were generated (Gross & Koch, 1991a). The sample shoots were kept in a temperature regulated pressure chamber during the entire experiment (Gross & Pham-Nguyen, 1987) and were dehydrated through stepwise increases in pressure. From the water potential isotherms, the bulk osmotic pressure at full turgor (π_o), and the bulk osmotic pressure when turgor initially reaches zero (π_p), the amount of symplastic (W_o) and apoplastic (W_a) water was determined. Before and after generating the water potential isotherms, 10 needles were removed at random from the shoot to determine their water content and to calculate the water content of the whole shoot needle mass. Thereafter they were used for the determination of the freezing point of their press sap by capillary microcryoscopy. By the end of the experiment the dry weight of needles still on the shoot was determined.

Capillary microcryoscopy of the needle press sap

The procedure for removing sap from the needles was described by Gross and Koch (1991b). Only the most important aspects will be described here. In order to avoid evaporation loss, directly after removal of the needles from the shoots, needles were submerged in paraffin oil, frozen in liquid nitrogen and then stored on dry ice. For the collection of the sap from single needles, a machine was developed (Walz Co., Effeltrich, FRG) which basically consists of two electrically operated plastic rollers. The thawed needles were squeezed between these rollers with 10 MPa pressure and the sap was continually collected with a capillary. Even small amounts of the sap from the small shade needles provide ample material. The preparation of the sap was always under paraffin oil.

The microcryoscope

The microcryoscope used in this study is a modern device developed by the second author (W. Koch) based on an experimental set up described by Kesseler (1957). The device consists of six parts: a cooling unit with bath, observation glass (stereo magnifying glass with zoom enlargement), lighting unit (xenon lamp, type 307, Leitz), external cryostat, steering device and automatic photographic registering device. A high precision cryostat guarantees a temperature adjustment within 0.01 °C. The electronical control unit allows variable timing in the adjustment of the temperature rise in the bath solution by broad changes in the heat flow (0.01 °C s^{-1} - 0.01 °C min^{-1}).

   The capillary microcryoscopic method is based on the following principle: ice crystals turn the plane of polarized light, thus becoming visible against a dark background when viewed through an "analyser" which has a polarized plane perpendicular to the oscillation plane of the "polarizer". Through this procedure, even the smallest ice particles, which are otherwise not recognizable in non-polarized light, are visible in the cryoscope capillaries.

Measurement procedures

The filled cryoscopic capillaries were attached to the frame of the capillary container, which was then placed in the cooling bath and sealed. The lamp was activated, and by turning the analyser, a dark field was produced in which only the frozen crystals of samples emitted light. The initial temperature of the bath fluid was always -40 °C. The temperature at which the ice crystals melt ("ice star" fading) represents the freezing point depression.

Calculation of the osmotic pressure

Using the method described, it was possible to determine 100 freezing point depressions by using the needle press sap from 10 needles. The average of these values served for calculating the osmotic pressure. Using tables, an osmolality (osmol per kg of water) was assigned to the respective freezing point depression. The osmotic pressure was then calculated based on the Van't Hoff relation $\pi = RTN/W$ at 25 °C. These values could then be compared with the results of the water potential isotherms also generated at 25 °C (cf. Gross and Koch, 1991a).

Determination of the apoplastic (W_{an}) and the symplastic water (W_{on}) as well as the bulk osmotic pressure (π_{on}) of the needles of a single shoot

The determination of the above mentioned parameters was achieved using the diagram $1/\pi$ vs. W, which allows a linear presentation of the relation between osmotic pressure (π) and water content (W) in a system (Fig. 1). (In this diagram the water content increases from right to left.)

The following values are needed for determination of π_{on}, W_{on} and W_{an}:

(i) the water content of all needles of a shoot at full water saturation (W_{tn}, distance CD) and at the moment of the final water potential measurement during the whole shoot water potential isotherm generation (distance CE). Both parameters were determined by weighing and they are plotted on the X-axis;

(ii) for each value of water content, the corresponding osmotic values of the needle press sap, determined by capillary microcryoscopy. These are the values of π_{sat} and π_{end}. The reciprocals of the π-values are point A $(1/\pi_{sat})$ and point B $(1/\pi_{end})$. (π_{sat} = osmotic pressure of the needle press sap at full water saturation; π_{end} = osmotic pressure of needle press sap at the end of the water potential isotherm generation.) The points A,B,C fall on a straight line, which represents the dependence of the osmotic pressure of the needle press sap on the needle water content.

The goal to determine π_{on} and W_{an} can be easily reached if a straight line can be drawn which represents the dependence of the bulk osmotic pressure of the needle symplast (π_n) on the symplastic water content (W_n). The construction of this line will depend on the following facts:

(i) the needle sap represents the sap of the vacuoles diluted by apoplastic water. Therefore, the values of the bulk osmotic pressure (π_n) must be lower than the π-values of the needle press sap;

(ii) under the assumption that the apoplastic water is almost pure (Tyree, 1976), the line should be shifted parallel to ABC and the determination of only one point should suffice for its construction.

This value is obtained by the PV-technique. The last water potential value of the whole shoot water potential isotherm would be most convenient. As of all measuring points of a water potential isotherm, the last point represents a balance pressure common to all tissue types (needles and wood) of a shoot and therefore can be used for needles alone. This point is found when the reciprocal value of the final balance pressure is plotted above the corresponding needle water content (point B').

A straight line across point B' is drawn in parallel to line AC. It intercepts the X-axis at point C' thus dividing total water content of the needles into symplastic (W_{on}) and apoplastic (W_{an}) water according to the theory of Tyree and Hammel (1972). The intercept with the Y-axis at point A' represents the reciprocal value of the bulk osmotic pressure at full water saturation (π_{on}). With the direct determination of the values A,B,C and B' in the context of the diagram shown in Fig. 1, a simple method for determining the bulk osmotic pressure, and the symplastic and apoplastic water content of the needles is obtained.

Capillary microcryoscopic measurements of the xylem fluid of 20 spruce shoots from a sample tree obtained using a pressure chamber yielded values between 0.05-0.28 MPa with an average of 0.105 MPa. For further calculations, a mean apoplastic osmotic pressure $\pi_a=0.1$ MPa (0.04 osmol kg^{-1}) was used, thereby increasing the value of W_{an}

approximately 3% and slightly decreasing the bulk osmotic pressure π_{on} of the needles. The dashed line in Fig. 1 illustrates this correction (A"B'C").

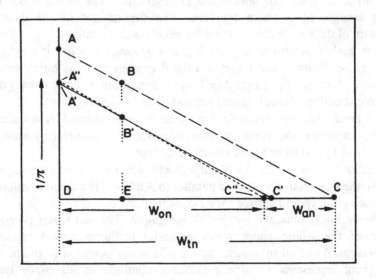

FIG. 1. Diagram for the determination of needle apoplastic water content (W_{an}) and the bulk osmotic pressure in the needle symplast (π_{on}) based on combined measurements of needle press sap capillary microcryoscopy and pressure-volume analysis of the whole shoot. ABC: line representing the dependence of the press sap osmotic pressure on the needle water content; A'B'C': line representing the dependence of the bulk osmotic pressure of the needle symplast on symplastic needle water content when apoplastic water is pure; A"B'C": the same relation as before when the osmolality of needle apoplastic solution equals 0.04 osmol kg[-1]. Note that water content increases from right to left. For further explanations see text.

In some experiments, the points A, B and C did not fall on a straight line. In order to compensate for the small deviations, the reciprocal values of the microcryoscopically measured osmotic pressures of the needle press sap represented by points A and B were averaged and a line was drawn from point C to halfway between A and B.

By subtracting π_{on}, W_{on} and W_{an} from the corresponding values π_o, W_o and W_a obtained from the PV analysis of whole shoots reported by Gross and Koch (1991a), it was possible to calculate the osmotic pressure at full water saturation (π_{os}), and the symplastic (W_{os}) and the apoplastic water content (W_{as}) of defoliated shoots. The following formulas were used:

$$\pi_o = \pi_{on} + \pi_{os} \qquad \text{Equation 1}$$

$$\pi_o = RT(N_{on} + N_{os}) (W_{on} + W_{os})^{-1} \qquad \text{Equation 2}$$

where N_{on} and N_{os} are the number of milliosmoles in the symplast of needles and defoliated shoots, respectively.

RESULTS

Apoplastic water

As seen in Table 1, the proportion of W_{an} on the total needle water content was as expected much less (13.8%) than the average proportion of W_{as} on the total water content of the defoliated shoots (26.7%). This is due to the greater proportion of xylem tissue in the defoliated shoot (Cheung, Tyree & Dainty, 1975).

TABLE 1. Apoplastic water of the current year's needles (W_{an}) and defoliated shoots (W_{as}) as a percentage of the corresponding total water content of the needles (W_{tn}) and shoots (W_{ts}) at full turgor. Averages, with standard errors in parenthesis, for 9 samples.

W_{an}/W_{tn} (%)	W_{as}/W_{ts} (%)
13.8 (1.7)	26.7 (3.8)

Bulk osmotic pressure

The bulk osmotic pressures given in Table 2 show that π_{on} was always 0.1 MPa lower than the values recorded by the PV analysis. The values of the bulk osmotic pressure in the symplast of whole shoots (π_o) depended on the contribution of needle and shoot mass and varied between π_{on} and the much higher π_{os}.

TABLE 2. Bulk osmotic pressures (MPa) at full turgor in the symplast of shoots (π_o), needles (π_{on}) and defoliated shoots (π_{os}) as well as the osmotic presssure of the needle press sap at full turgor (π_{sat}). Averages, with standard errors in parenthesis, for 9 samples.

π_o	π_{on}	π_{os}	π_{sat}
-1.80 (0.06)	-1.93 (0.05)	-1.51 (0.14)	-1.7 (0.06)

CONCLUSIONS

The combined use of pressure-volume analysis and capillary microcryoscopy offers a reliable possibility for determining the apoplastic water content and other water relations parameters in conifer needles with high accuracy. The new procedure of capillary microcryoscopy is essential for the determination of osmotic values of extremely small press sap quantities (0.1 nl) from conifer needles. To reach results with sufficient accuracy, the following points should be considered:
(i) During the water potential isotherm generation the equilibration of water in the tissue of sample shoots must be guaranteed. Inevitably this takes much time. Tyree et al. (1978) did show that pressure-volume determinations which did not allow sufficient time for establishment of the water status equilibrium of the tissue in between measurements led to overestimation of the bulk osmotic pressure and apoplastic water content. According to Ritchie & Shula (1984), who worked with Pseudotsuga menziesii seedlings, equilibrium is only partly reached after more than 70 minutes. The lack of equilibration of water in the tissue may have two consequences: first, the water potential values of the whole sample shoots are incorrect and second, the microcryoscopically determined π-values of the sap of each needle could differ greatly. Such results are unsuitable for calculation of needle apoplastic water content and other parameters;
(ii) some uncertainties exist about the determination of the osmolality of the apoplastic solution. It is not clear whether the osmolality of the xylem sap used here reflects the actual situation in the apoplast. In addition, there are probably variations in the osmolality of the xylem sap obtained

from shoots with different water contents. However, the errors resulting from such uncertainities are probably small;
(c) the proper use of the capillary microscope requires a lot of practical experience.

In summary, the method described here for determining the amount of apoplastic water and other water relations parameters in conifer needles is well suited for refined laboratory scale experiments.

REFERENCES

Cheung, Y.N.A., Tyree, M.T. & Dainty, J. (1975) Water relation parameters on single leaves obtained in a pressure bomb and some ecological interpretations. *Canadian Journal of Botany*, **53**, 1342-1346.

Gross, K. & Koch, W. (1991a) Water relations of *Picea abies*.I. Comparison of water relations parameters of spruce twigs at the end of the vegetation period and in winter. *Physiologia Plantarum*, **83**, 290-295.

Gross, K. & Koch, W. (1991b) Water relations of *Picea abies*. II. Determination of the apoplastic water content and other water relations parameters of spruce needles by means of microcryoscopy and the pressure-volume analysis. *Physiologia Plantarum*, **83**, 296-303.

Gross, K. & Pham-Nguyen, T. (1987) Pressure-volume analyses on shoots of *Picea abies* and the leaves of *Coffea liberica* at various temperatures. *Physiologia Plantarum*, **70**, 189-195.

Kesseler, H. (1957) Mikrokryoskopische Unter-suchungen an Meeresalgen mit besonderer Berücksichtigung der Turgorregulations-vorgänge von *Chaetomorpha linum*. Thesis, University of Kiel, FRG.

Ritchie, G.A. & Schula R.G. (1984) Seasonal changes of tissue water relations in shoots and root systems of Douglas-fir seedlings. *Forest Science*, **30**, 538-548.

Tyree, M.T. (1976) Negative turgor pressure in plant cells: fact or fallacy? *Canadian Journal of Botany*, **54**, 2738-2746.

Tyree, M.T. & Hammel (1972) The measurement of the turgor pressure and the water relations of plants by the pressure-bomb technique. *Journal of Experimental Botany*, **23**, 267-282.

Tyree, M.T., MacGregor, M.E., Petrov, A. & Upenieks, M.I. (1978) A comparison of systematic errors between the Richards and Hammels methods of measuring tissue-water relations parameters. *Canadian Journal of Botany*, **56**, 2153-2161.

The assessment of water status in chilled plants

A. PARDOSSI , P. VERNIERI AND F. TOGNONI

Dipartimento di Biologia delle Piante Agrarie
Sezione di Orticoltura e Floricoltura, Università degli Studi di Pisa
Viale delle Piagge, 23, 56100 Pisa, Italy

SUMMARY

A study was carried out to compare the pressure chamber and thermocouple psychrometer method for the assessment of leaf water potential in tomato, eggplant and bean plants under chilling temperatures. The results reported here and the examination of the literature on water relations of chilled plants indicate that great care is necessary when using the pressure chamber method to evaluate the water status of plants under chilling temperatures.

INTRODUCTION

Plant water status is usually assessed by measuring the water potential with the pressure chamber (PC) or the thermocouple psychrometer (TP) (Campbell, 1985). In many species the water potentials determined with the PC ($\Psi_w{}^{PC}$) and TP ($\Psi_w{}^{TP}$) have shown good correspondence (Ritchie & Hinckley, 1975; Campbell, 1985). Nevertheless, several authors have reported discrepancies between $\Psi_w{}^{PC}$ and $\Psi_w{}^{TP}$, with lower values of $\Psi_w{}^{PC}$ at low potentials (Ritchie & Hinckley, 1975; Turner, 1986; Hardegree, 1989).

We report here the main results of a study carried out to evaluate the PC method for measuring leaf Ψ_w in both chilled and unchilled plants of three chilling-sensitive species.

MATERIALS AND METHODS

Different experiments were conducted with 5- to 10-week-old tomato (*Lycopersicon esculentum* Mill., cv. Marmande) and eggplant (*Solanum melongena* L., cv. Violetta Lunga di Romagna) seedlings with 8 to 10 leaves, and with 2- to 3-week-old bean (*Phaseolus vulgaris* L., cv. Mondragone) seedlings with fully-expanded primary leaves.

A range of leaf water status was achieved by subjecting seedlings to drought and chilling treatments of different magnitude and duration. Drought was induced by withholding water or by excising the plant above the root insertion. Chilling treatment was performed by transferring well-hydrated plants into a cold room (3 ± 0.5 °C) at approximately 70% RH in darkness or in continuous fluorescent light (15 W m^{-2}). Chilling lasted 3 to 72 h according to the experiment.

All measurements were made at ambient laboratory temperature on the 4th, 5th or 6th oldest leaves in tomato and eggplant and on the primary leaves in bean.

Ψ_w^{TP} values were determined on leaf segments (3 x 2 cm) using a SC-104 TP connected to a NT-3 nanovoltmeter (Decagon Devices, Pullman, WA, USA). The TP was calibrated with NaCl solutions. The operating procedures described by Welbaun & Bradford (1988) were observed.

Ψ_w^{PC} measurements were made according to standard procedure. The test leaf was excised and rapidly inserted into a humidified PC. Whole leaves were used in tomato.

To correct the PC readings for Ψ_s^{apo}, the osmotic potential of xylem sap was measured in control, droughted and chilled seedlings in some experiments. Xylem sap was obtained by subjecting plant tops (1.5 to 4 g fresh weight) to slight overpressure in the PC until sap flowed out of the cut surface. In tomato and eggplant, a short tygon tube was fitted to the shoot protruding through the bung, and a micropipette was used to collect sap. The first 40 to 60 µl of exudate were collected from each plant. Pooled fractions from three to five plants were assayed for osmotic potential by osmometry. In bean seedlings a filter paper disc was saturated with expressed sap and immediately inserted into the TP for measurement.

For the PC *vs.* TP comparison, the psychrometric determinations were made either on the same leaf used for Ψ_w^{PC} (after measurements) or on a different leaf excised from the same plant.

RESULTS

The PC data were corrected for the presence of solutes in the apoplast using the average Ψ_s^{apo}, determined for each species and for each stress treatment. $\Psi_s{}^{apo}$ ranged from -0.1 to -0.17 and was independent of stress conditions and Ψ_w.

In unchilled plants, $\Psi_w{}^{PC}$ agreed well with $\Psi_w{}^{TP}$ over the considered potential range (Fig. 1). On the other hand, quite higher discrepancies with higher $\Psi_w{}^{PC}$ values (up to 1.4 MPa) were found in chilled plants (Fig. 1).

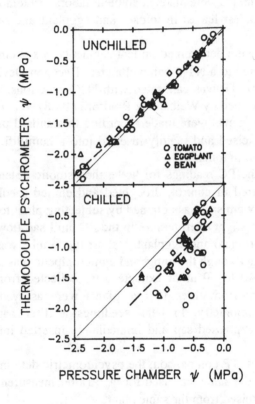

FIG. 1. Relationship between pressure chamber and thermocouple psychrometer measurements of leaf water potential in unchilled and chilled seedlings of tomato, eggplant and bean. Pressure chamber measurements were corrected for the apoplastic osmotic potential. The solid lines represent equipotential lines and the dashed lines give the fitted linear regressions. Correlation coefficients were 0.98 ($P < 0.001$) and 0.66 ($P < 0.01$) for unchilled and chilled plants, respectively. Individual points are means of 6 to 12 determinations and derive from different experiments.

Our determinations were made at 20-25 °C. Since the temperature can affect the PC measurements (Tyree, Benis & Dainty, 1973; Gross & Pham-Nguyen, 1987), we also compared the PC and TP methods at 3 °C in chilled and droughted plants as well as in plants previously droughted and then chilled for a few hours. The PC again overestimated Ψ_w in chilled plants, while in droughted or droughted-chilled plants $\Psi_w{}^{PC}$ and $\Psi_w{}^{TP}$ corresponded well (data not shown).

DISCUSSION

Our results indicate that the PC may overestimate the Ψ_w values in chilled plants. We are not aware of other papers on chilled plants reporting a comparison between PC and TP or indicating possible anomalies in measurements obtained using the PC method. Nevertheless, an examination of the literature dealing with Ψ_w of chilled plants throws further doubts on the accuracy of the PC in assessing the water status in chilling-sensitive species under chilling temperatures. Determining Ψ_w by psychrometry, Markhart et al. (1980), McWilliam, Kramer & Musser (1982), Eamus, Fenton & Wilson (1983), Eamus (1986, 1987a, 1987b) and we also, Vernieri, Pardossi & Tognoni (1991), have observed significant declines in Ψ_w and, when measured, Ψ_s and Ψ_p. In contrast, no substantial decreases in Ψ_w have been reported in studies using the PC on chilling-sensitive plants (Daie & Campbell, 1981; Daie, Campbell & Seeley, 1981; Eze, Dumbroff & Thompson, 1983; Zhang, Li & Brenner, 1986; Flores et al., 1988): in these papers Ψ_w values appear questionable.

In studies on the effect of temperature on leaf ABA content, Daie and co-workers (Daie & Campbell, 1981; Daie et al., 1981) observed normal Ψ_w values in chilled plants of bean, tomato, eggplant, corn and okra. They concluded that ABA levels increased in response to low temperature per se and not to cold-induced water stress. However, other authors (Eamus & Wilson, 1983; Eze et al., 1983; Capell & Dorffling, 1989) as well as our group (Vernieri, Pardossi & Tognoni, 1989; Vernieri et al., 1991) have observed a rise in the ABA content of chilled plants only when moisture deficit occurred, and questioned the reliability of the water status determinations made by Daie and colleagues.

On the other hand, the conclusions of Eze et al. (1983) for chilled bean appear to be based on the determinations of RWC rather than on the PC readings, since the latter changed very slightly in chilled plants with respect to controls.

In studies on the influence of mefluidide on the chilling resistance of corn, Zhang *et al.* (1986) reported a small increase of Ψ_w and a large decrease of Ψ_s for untreated plants at 4 °C. A large rise in Ψ_p can be derived from these data, notwithstanding a drop in RWC as reported by the authors. This can be explained by a rather intense osmotic adjustment and/or a change in leaf elasticity, which are not expected to occur in relatively short periods of chilling, as suggested by observations made in this study (data not shown) or reported elsewhere (Eamus *et al.*, 1983; Vernieri *et al.*, 1989, 1991). Hence, Ψ_w values could have been overestimated by these authors.

Flores *et al.* (1988) observed a strong water deficit in chilled cucumber seedlings (fresh weight loss of leaves was nearly 75% of initial weight after 24 h chilling), but only slight changes in Ψ_w (around -0.25 MPa in chilled seedlings against -0.15 MPa in controls) were reported.

In the papers mentioned above, the authors did not report PC readings being corrected for $\Psi_s{}^{apo}$, and this could account for the apparent overestimate of Ψ_w.

We know of only a few papers reporting the same kind of PC *vs.* TP relationship for unchilled plants as that obtained with chilled plants in our study. In these papers the disparity between the two methods is simply explained (although only partially in some cases) by the non-correction of P for $\Psi_s{}^{apo}$ (Kauffmann, 1968, for Engelmann spruce; West & Gaff, 1971; Spomer & Langhans, 1972; Frank & Harris, 1973).

Decreases in $\Psi_s{}^{apo}$ can occur in plants as a consequence of contamination of the apoplastic solution by intracellular solutes from damaged or leaky cells (Kaplan & Gale, 1974; Campbell, 1985). Turner (1976) reported that membrane damage in plant leaves caused characteristic change in the PC measurements; in particular, the pressure required to first express sap in frozen/thawed leaves of oats was around 0.9 MPa lower than that required for unfrozen leaves. Since damage to membrane and electrolyte leakage often occurs in chilled tissues (Wolk & Herner, 1982), the resulting low $\Psi_s{}^{apo}$ values may induce an overestimate of Ψ_w by the PC method.

In our work, on the other hand, $\Psi_w{}^{PC}$ did not correspond to $\Psi_w{}^{TP}$ notwithstanding the correction for $\Psi_s{}^{apo}$. $\Psi_s{}^{apo}$ in chilled plants was rather high, although in some experiments (data not shown) as well as in other work (Vernieri, Pardossi & Tognoni, 1989) electrolyte leakage from chilled leaves was observed in tomato and bean. The amount of xylem sap collected to measure $\Psi_s{}^{apo}$ represented a small fraction of the total apoplastic water inside the plant tops, and, moreover, it was presumably released from the stem and the petiole and not from the leaf blade, according to Jachetta, Appleby & Boersma (1986). This could account for the high values of Ψ_s^{apo} we found in chilled plants. On the other hand, Turner (1976) concluded that membrane damage affects the PC measurements only when the majority of cells in the leaf are injured. As chilling often did not produce injury over the total leaf area, the disparity between PC and TP in chilled plants should be explained by other factors.

In particular, the disparity between the instruments may be due to the disequilibrium in the leaf water and the presence of large potential gradients across the blade. A gradient in Ψ_w can be expected since there is a high resistance to water flow within the leaf (Turner, 1986) and low temperatures decrease the leaf hydraulic conductivity (Tomos *et al.,* 1981; Eamus & Wilson, 1984; Gross & Pham-Nguyen, 1987). The slow attainment of equilibrium within the leaf during a PC measurement can induce overestimates of leaf Ψ_w, as PC determines Ψ_w of the wettest zone of the leaf (Meiri, Plaut & Shimshi, 1975; Turner, Spurway & Schulze, 1984).

Our results confirm that the water potential measurements made with the pressure chamber on herbaceous plants should be handled with care and that a better assessment of plant water status may even be achieved by measuring other parameters such as relative water content and osmotic and turgor potential.

AKNOWLEDGMENTS

This research was supported by National Research Council of Italy, Special Project RAISA, Sub-project No. 2.

REFERENCES

Campbell, G.S. (1985) Instruments for measuring plant water potential and its components. In *Instrumentation for Environmental Physiology* (ed. B. Marshall and F.I. Woodward), pp. 193-214. Cambridge University Press, Cambridge.

Capell, B. & Dorffling, K. (1989) Low temperature-induced changes of abscisic acid contents in barley and cucumber leaves in relation to their water status. *Journal of Plant Physiology,* **135,** 571-575.

Daie, J. & Campbell, W.F. (1981) Response of tomato plants to stressful temperatures. *Plant Physiology,* **67,** 26-29.

Daie, J., Campbell, W.F. & Seeley, S.D. (1981) Temperature-stress-induced production of abscisic acid and dihydrophaseic acid in warm- and cool-season crops. *Journal of American Society for Horticultural Science,* **106,** 11-13.

Eamus, D. (1986) The responses of leaf water potential and leaf diffusive resistance to abscisic acid, water stress and low temperature in *Hibiscus esculentus*: the effect of water stress and ABA pre-treatments. *Journal of Experimental Botany,* **37,** 1854-1862.

Eamus, D. (1987a) Influence of preconditioning upon the changes in leaf conductance and leaf water potential of soybean, induced by chilling, water stress and abscisic acid. *Australian Journal of Plant Physiology,* **14,** 331-339.

Eamus, D. (1987b) Stomatal behaviour and leaf water potential of chilled and water-stressed *Solanum melongena*, as influenced by growth history. *Plant, Cell and Environment,* **10,** 649-654.

Eamus, D., Fenton, R. & Wilson, J.M. (1983) Stomatal behaviour and water relations of chilled *Phaseolus vulgaris* L. and *Pisum sativum* L. *Journal of Experimental Botany,* **34,** 434-441.

Eamus, D. & Wilson, J.M. (1983). ABA levels and effects in chilled and hardened *Phaseolus vulgaris. Journal of Experimental Botany,* **34,** 1000-1006.

Eamus, D. & Wilson, J.M. (1984) The effect of chilling temperatures on the water relation of leaf epidermal cells of *Rhoeo discolor. Plant Science Letters,* **37,** 101-104.

Eze, J.M.O., Dumbroff, E.B. & Thompson, J.E. (1983) Effects of temperature and moisture stress on the accumulation of abscisic acid in bean. *Physiologia Plantarum,* **58,** 179-183.

Flores A., Grau A., Laurich, F. & Dorffling, K. (1988) Effect of new terpenoid analogues of abscisic acid on chilling and freezing resistance. *Journal of Plant Physiology,* **132,** 362-369.

Frank, A.B. & Harris, D.G. (1973) Measurement of leaf water potential in wheat with a pressure chamber. *Agronomy Journal,* **65,** 334-335.

Gross, K. & Pham-Nguyen, T. (1987) Pressure-volume analyses on shoots of *Picea abies* and leaves of *Coffea liberica* at various temperatures. *Physiologia Plantarum,* **70,** 189-195.

Hardegree, S.P. (1989) Xylem water holding capacity as a source of error in water potential estimates made with the pressure chamber and thermocouple psychrometer. *American Journal of Botany,* **76,** 356-360.

Jachetta, J.J., Appleby, A.P. & Boersma, L. (1986) Use of the pressure vessel to measure concentrations of solutes in apoplastic and membrane-filtered symplastic sap in sunflower leaves. *Plant Physiology,* **82,** 995-999.

Kaplan, A., & Gale, J. (1974) Modification of the pressure bomb technique for measurement of osmotic potential in halophytes. *Journal of Experimental Botany,* **25,** 663-668.

Kaufmann, M.R. (1968) Evaluation of the pressure chamber technique for estimating plant water potential of forest tree species. *Forest Science,* **14,** 369-374.

Markhart III, A.H., Peet, M.M., Sionit, N. & Kramer, P.J. (1980) Low temperature acclimation of root fatty acid composition, leaf water potential, gas exchange and growth of soybean seedlings. *Plant, Cell and Environment,* **3,** 435-441.

McWilliam, J.R., Kramer, P.J., Musser, R.L. (1982) Temperature-induced water stress in chilling-sensitive plants. *Australian Journal of Plant Physiology*, **9**, 343-352.

Meiri, A., Plaut, Z. & Shimshi, D. (1975) The use of the pressure chamber technique for measurement of the water potential of transpiring plant organs. *Physiologia Plantarum*, **35**, 72-76.

Ritchie, G.A. & Hinckley, T.M. (1975) The pressure chamber as an instrument for ecological research. *Advances in Ecological Research*, **9**, 165-254.

Spomer, L.A. & Langhans, R.W. (1972) Evaluation of pressure bomb and dye method measurements of tissue water potential in greenhouse chrysanthemum. *Hort. Science*, **7**, 412-413.

Tomos, A.D., Steudle, E., Zimmermann, U. & Schulze, E.D. (1981) Water relations of leaf epidermal cells of *Tradescantia virginiana*. *Plant Physiology*, **68**, 1135-1143.

Turner, N.C. (1976) Use of the pressure chamber in membrane damage studies. *Journal of Experimental Botany*, **27**, 1085-1092.

Turner, N.C. (1986) Crop water deficits: a decade of progress. *Advances in Agronomy*, **39**, 1-51.

Turner, N.C., Spurway, R.A. & Schulze, E.D. (1984) Comparison of water potentials measured by *in situ* psychrometry and pressure chamber in morphologically different species. *Plant Physiology*, **74**, 316-319.

Tyree, M.T., Benis, M. & Dainty, J. (1973) The water relations of hemlock (*Tsuga canadensis*). III. The temperature dependence of water exchange in a pressure bomb. *Canadian Journal of Botany*, **51**, 1537-1543.

Vernieri, P., Pardossi, A. & Tognoni, F. (1989) Chilling-induced water stress in tomato: effect on abscisic acid accumulation. *Advances in Horticultural Science*, **2**, 78-80.

Vernieri, P., Pardossi, A. & Tognoni, F. (1991) Influence of chilling and drought on water relations and abscisic acid accumulation in bean. *Australian Journal of Plant Physiology*, **18**, 25-35.

Welbaum, G.E. & Bradford, K.J. (1988) Water relations of seed development and germination in muskmelon (*Cucumis melo* L.). *Plant Physiology*, **86**, 406-411.

West, D.W. & Gaff, D.F. (1971) An error in the calibration of xylem-water potential against leaf-water potential. *Journal of Experimental Botany*, **22**, 342-346.

Wolk, W.D., & Herner, R.C. (1982). Chilling injury of germinating seeds and seedlings. *HortScience*, **17**, 169-173.

Zhang, C.L., Li, P.H. & Brenner, M.L. (1986) Relationship between mefluidide treatment and abscisic acid metabolism in chilled corn leaves. *Plant Physiology*, **81**, 699-701.

An artificial osmotic cell : a model system for studying phenomena of negative pressure and for determining concentrations of solutes

H. HEYDT AND E. STEUDLE

Universität Bayreuth, Universitätsstraße 30, D-8580 Bayreuth, Germany.

INTRODUCTION

For technical reasons, it is very difficult to measure negative pressures in the xylem of transpiring plants directly. An artificial osmotic cell has been constructed using reverse osmosis membranes to study phenomena of negative pressure (Steudle & Heydt, 1988; Zhu, Steudle & Beck, 1989). The cell has been also used as an analytical device (osmotic sensor; Steudle & Stumpf, 1989) to measure the concentration of certain solutes in solutions. The sensivity and selectivity of the osmometer could be increased by coupling the osmotic process to a chemical reaction.

MATERIAL AND METHODS

Reverse osmosis membranes were used to separate the cell interior from the medium. 'Cell turgor' was measured with the aid of a pressure transducer and was recorded continuously. The system was calibrated in the range of positive and negative pressures by applying gas pressures to both sides of the transducer membrane. During the experiments, the calibration and the proper function of the transducer could be checked (Steudle & Heydt, 1988). The thickness of the osmotic cell was 0.1 to 0.2 mm and the diameter 4 mm. The cell was filled with a non-permeating solute to build up a positive turgor in the presence of hypotonic solutions. In order to create negative pressures (tensions), the external medium was replaced by hypertonic solutions of non-permeating solutes.

RESULTS AND DISCUSSION

When permeating solutes were added to the medium, biphasic changes of turgor were observed. From these pressure-time curves, the hydraulic conductivity (L_p), the permeability (P_s), and the reflection (σ_s) coefficients of the membrane were evaluated. It could be shown that the transition from positive to negative pressures did not affect the absolute values of the coefficients P_s, σ_s, and L_p and, hence, the transport properties of the membrane.

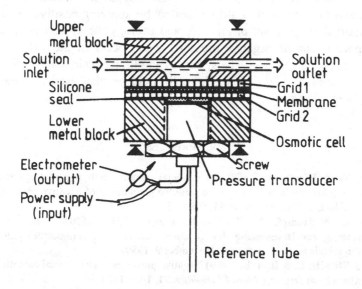

FIG. 1. The experimental set up. An osmotic cell was separated by a membrane from the solution. The membrane was supported at both sides by a grid so that positive as well as negative pressures could be built up in the cell. The pressure in the cell was recorded continuously by a pressure transducer and its function could be checked during the experiments by applying known pressures to the reference tube and observing the response.

The responses in pressure were proportional to the concentrations applied. Concentrations of both, permeating and non-permeating solutes could be determined in a single experiment. Chemical reactions in the osmotic cell were used to improve the selectivity and sensivity of the system. In one experiment, the cell was filled with water and solid silver chromate (Ag_2CrO_4) to determine ammonia. NH_3 was rapidly permeating the membrane and dissolved the Ag_2CrO_4 (Ag_2CrO_4 + 4NH_3 = 2$Ag(NH_3)_2$ + + CrO_4 2-). The solutes created in the cell increased the osmotic pressure and the 'cell turgor'. The rate of increase was proportional to the concentration of NH_3 of the medium. Sensivity: 10^{-4} to 10^{-5} M.

The artificial osmotic cell will be employed to determine concentrations of solutes in mixtures selectively. The apparatus can be applied to measure and control concentrations in biotechnology and in medicine and during chemical processes. Negative pressures in an artificial osmotic cell of up to -0.7 MPa (absolute) were created in short and -0.3 MPa (absolute) in long terms. Cavitations which occurred spontaneously in the cell could be 'healed' by applying positive pressure for a short time. In plant physiology and ecology the system is used to study phenomena of negative pressure which occur in the xylem and, perhaps, in living cells too. The system is employed to simulate water and solute relations of plant cells. The cell could be also used to measure water potential of soils.

REFERENCES

Steudle, E. & Heydt, H. (1988) An artificial osmotic cell: a model system for simulating osmotic processes and for studying phenomena of negative pressures in plants. *Plant, Cell Environment*, 11, 629-637.

Steudle, E. & Stumpf, B. (1989) German patent DE 3825208. Verfahren und Einrichtung zur Bestimmung des Gehaltes von einem Lösungsmittel gelößten Stoffen mittels eines Osmometers. November 2, 1989.

Zhu, J.J. Steudle, E. & Beck E. (1989) Negative pressures in an artificial osmotic cell by extracellular freezing. *Plant Physiology*, 91, 1454-1459.

Measurement of water and solute uptake into exised roots at positive and negative root pressures

H. HEYDT AND E. STEUDLE

Universität Bayreuth, Universitätsstraße 30, D-8580 Bayreuth, Germany.

INTRODUCTION

Transport properties of roots have been characterized by certain transport coefficients such as the hydraulic conductivity (Lp_r), the permeability (P_{sr}), and reflection coefficient (σ_{sr}). So far, these coefficients have been measured only at positive root pressures. However, according to the cohesion theory of the ascent of sap, negative pressures exist in the xylem of transpiring plants. Therefore, it is of some interest whether or not the absolute values of the transport coefficients would change when pressures in the xylem drop to negative values, *i.e.* below vacuum (= 0 MPa).

In the present work, Lp_r, P_{sr}, and σ_{sr} of excised roots of *Zea mays* L. have been measured at positive and negative root pressures with the aid of the root pressure probe. Furthermore, responses of xylem pressures of intact maize plants to changes in the rate of transpiration have been followed directly.

MATERIALS AND METHODS

The hydrostatic hydraulic conductivity (Lp_{rh}) of root endsegments was determined with the aid of a root pressure probe. In 'hydrostatic experiments' a metal rod was moved into (out of) the probe and hence, root pressure was changed. Lp_{rh} was calculated from the following relaxation of root pressure (Steudle, Oren & Schulze, 1987). Alternatively, in 'osmotic experiments' the concentration of a solute in the medium was changed and Lp_r, P_{sr}, and σ_{sr} were determined from the pressure responses (Steudle *et al.*, 1987). The addition of a non-permeating solute (mannitol) caused the root pressure to drop to a stationary negative pressure. Now, a permeating solute (Ethanol, $NaNO_3$) was added or removed from the medium and again, Lp_r, P_{sr}, and σ_{sr} were calculated from root pressure responses. For measurements in the range of negative pressures, all air spaces within the cortex had to be infiltrated with water to avoid cavitations during the measurement. This was done by pressurizing excised roots for 10 min. prior to experiments.

Measurements have been also performed on whole plants grown on vermiculite. Root tips grown out of the plastic pots were connected to a root pressure probe while the shoot was covered with a plastic bag to create positive root pressures.

RESULTS AND DISCUSSION

By comparing data obtained for infiltrated roots with those for non-infiltrated ones, it was shown that infiltration did not affect the transport properties of the excised roots. At levels of negative pressures, no hydrostatic experiments could be performed because of cavitations which occured at the surface of the metal rod when root pressure was manipulated, but it was possible to carry out osmotic experiments at negative pressures. No changes of the absolute values of Lp_r, P_{sr}, and σ_{sr} could be detected when root pressure dropped from positive (+0.3 MPa) to negative (-0.2 MPa) root pressures.

On the experiments with intact plants the plastic bag was removed after a constant root pressure was established. In this way a decrease in root pressure was induced (response time less than a minute). It was not possible to measure negative pressures in these experiments because the roots of intact plants could not be infiltrated so embolism occurred before root pressure reached 0 Mpa (vacuum). The changes were reversible, *i.e.* root pressures increased again when the shoot was covered with a plastic bag before an embolism was observed. The quick responses of root pressure to transpiration showed that there was a good hydraulic contact between root and shoot.

REFERENCES

Heydt, H. & Steudle, E. (1991) Measurement of negative pressure in the xylem of excised roots: Effects on water and solute relations. *Planta,* **184,** 389-396.
Steudle, E., Oren, R. & Schulze, E.D. (1987) Water transport in maize roots. *Plant Physiology,* **84,** 1220-1232.

Index